Hartmut Polzin

Inorganic Binders
for mould and core production in the foundry

For Elke, Fritz, Clara, Luise and Moritz

Hartmut Polzin

Inorganic Binders

for mould and core production in the foundry

SCHIELE&SCHÖN

ISBN: 978-3-7949-0884-4

© Fachverlag Schiele und Schön GmbH 2014
1. edition

This book is based on the habilitation thesis of Dr. Hartmut Polzin
at the TU Bergakademie Freiberg, june 2012

cover picture: Pinter Guss, Deggendorf

This work is subject to copyright. All rights are reserved, whether the whole or part of the material is concerned, specifically the rights of translation, reprinting., reuse if illustrations, recitation, broadcasting. reproduction on microfilms or in any other ways, and storage in data banks. Duplication of this publication or parts thereof is only permitted under the provisions of the German Copyright Law of September 9. 1965, in its current version, and permission for use must always be obtained from Fachverlag Schiele & Schön GmbH are liable to prosecution under the German Copyright Law.
The use of general descriptive names, registered names, trademarks, etc. in this publication does not imply, even in the absence of a specific statement, that such names are exempt from the relevant protection laws and regulations and therefore free for general use.

© 2014 Fachverlag Schiele & Schön GmbH, Markgrafenstr. 11, 10969 Berlin.
All rights reserved, in particular the translation in other languages.
Without the prior written permission of the above publisher of this book,
no part of the publication may be repruduced.

Editorial Office Bernhard E. Schwarz
Layout Karen Weirich/Eva Hernandez
Composing Fachverlag Schiele & Schön GmbH
Production CPI Clausen & Bosse, Leck, printed in Germany

Foreword

In addition to clay minerals, which have been used for decades as a binder for the compaction moulding process (with bentonite moulding materials), there are also inorganic chemically curing binder systems with a long tradition in the foundry. Since the Forties of the last century cement has been used for mould and core production. The core production was then revolutionized in the Fifties by the water glass CO_2 process. Due to the short cycle times this cold box process became very popular. Introduced into practice in the Fifties, the classical inorganic systems superseded the organic binder systems increasingly around the 70s and 80s.

The main reasons for this development were the higher performance, the high process reliability, and improved technological properties. Due to a constantly increasing environmental awareness in the foundry industry, which has been increasingly underpinned by government calls for an improvement in casting production, the almost forgotten inorganic binder systems had a renaissance at the turn of the century.

In addition to two conferences on the subject in 2002 in Wuppertal and 2005 in Hannover, interested experts at GIFA 2003 were presented a series of inorganic binder systems. Entirely new at this exhibition were the two salt binder systems Hydrobond and Laempe-Kuhs. The second group of the presented binder systems were the silicate-based products. This group of binder systems were based mainly on silicate binder components belonging to the classic water glass binder. To compensate the well-known disadvantages of water glass binder, the systems typically operate with additives and adjuvants, contained either in the binder or added as a liquid or powder component in the moulding material preparation.

The initiated trend towards the increased use of inorganic binder systems and its produced moulding mixtures is not yet to be questioned and needs to be further pursued in the coming years. There is some misleading and conflicting information about the current state of the application as well as its achievable property level.

Therefore, this book should be attempted as complete as possible. It is aiming to provide answers to the question of what can afford inorganic binder systems at the present. On the other hand this book should open questions or problems to be solved for a further increase in proportions in the mould and core production over the coming years.

Without the support of a number of people this book would not have been possible. I want to thank all colleagues who have supported me through the provision of image and text material. Many of the results presented here were obtained by students at the Foundry-Institute of TU Bergakademie Freiberg through diligent work in various research papers. My thanks at this point to Christoph Birnbaum, Sebastian Marx, Madlen Nicklisch, Marcel Nürnberg, Ronny Reuther, Axel Wezel and Martin Wrase. I would also like to thank Frank Gleißner, whose support was tireless and reliable when designing or editing images and graphics.

When talking about inorganic binder systems, I would very much like to remember to Professor Eckart Flemming (1929–2004), who worked his whole life as a researcher on the water glass moulding process and would meet the current development of "the inorganic" with great joy. To him I am obliged to great thanks and if this book dedicates to him, hence.

Hartmut Polzin
Freiberg, April 2012

Contents

	Forword	5
1	The Beginnings of the Application of Inorganic Binder Systems	9
2	The Development and State of the Application up to the Year 2000	11
3	Overview of Currently Available Inorganic Binder Systems	14
3.1	Alkali silicate binder (water glass binder)	14
3.2	Silica sol for investment casting	31
3.3	Cement as a binder system	32
3.4	Geopolymer binder	36
3.5	Salt binder systems	37
3.6	Gypsum as binder	39
4	Classification of Moulding Processes with Inorganic Binder Systems	42
4.1	Hardening by gasification process	43
4.1.1	Water glass CO_2 process	43
4.1.2	Water glass warm air process	61
4.1.3	Warm air drying in water glass powder systems	65
4.2	Cold self-curing processes	69
4.2.1	Cement moulding process	72
4.2.2	Water glass ester process	89
4.2.3	Geopolymer process	108
4.3	Warm or hot curing processes	114
4.3.1	Processses with tempered mould tools	114

4.3.2	Microwave drying method	121
4.3.3	Methods with salt binder systems	132
4.3.4	Investment casting with silica sol binders	137
5	**The Use of Alternative Moulding Materials**	**146**
6	**Reclamation of Used Sands**	**163**
7	**The Influence of Inorganic Binder on Clay Bonded Circulation Moulding Materials**	**202**
	Index	211

1 The Beginnings of the Application of Inorganic Binder Systems

Inorganic binder systems such as loam or clay were principally used in foundries since their beginnings around 5000 years ago. But, if one looks at the group of chemical curing systems, the application period is significantly limited. Probably the oldest chemically (inorganic) curing and moulding material binder system is cement. The first field tests were performed with cement according to *Roll* [1.1] at the turn of the century in 1900. The hydraulic cement binder however, gained practical importance only through the work of *Durand*. In Germany *Goedel* was the first to deal with the process. Application of the process was mainly used in the production of castings made of cast steel. An early work, which deals with the basics of the cement moulding material process is found in [1.2].

Another group of inorganic binder systems that have long been used in the foundry industry are the silicic acid and silicate binder solutions. Silica sols are solutions of silicon dioxide in water and applied as a binder in the lost wax or investment casting, and in a number of so-called precision casting procedures. *Hinz* deals with the basics of these hardening by drying binder systems in [1.3]. Alkali silicate solutions, better known as water glass solutions, have been used in foundries since around 1950. *Petrzela* introduced the water glass process to the public with a patent in 1947 [1.4] [1.5]. In this way the first 'cold box' process was available and revolutionized core production in particular, bringing significantly shorter curing times. Working almost in parallel, Ljass's work also led to the development of gas hardening in the water glass process [1.6] [1.7]. This process is still used today though in relatively small scale for the production of cores in all areas of cast material.

After the water glass CO_2 process had provided a significant advance in productivity and process reliability of casting production, a number of other curing technologies for the water glass binder in the cold self-curing process were developed in the subsequent period. During the water glass powder curing process [1.8] [1.9] dicalcium and tricalcium silicates were used as a main component in Portland cement as a powder hardener. The process represented a further development of the water glass-silicide process (Nishiyama process) that was used with powdered ferrosilicon as a hardener [1.10] [1.11]. A further variant of the process was the self-curing water glass clay process in which a plastic moulding material is solidified by a combination of compression and chemical curing. By the addition of bentonite or clay, the moulding material produced may be separated immediately after the compaction of the core box.

The most important form of self-curing technology with regard to water glass binder was, and still is, the water glass ester process [1.12] [1.13]. This was used to achieve the solidification reactions in raw moulding material organic esters on the basis of acetic acid or propylene carbonate, for example. The process is applied today for the production of moulding materials and cores mainly in the area of non-ferrous castings manufacturing in addition to the cement moulding material process which is the only significant cold self-curing inorganic moulding process.

In the course of development, other inorganic components were tested as moulding material plastic binders but these will not be explored in this discussion at this point. Noteworthy however, is gypsum which is used even today in a number of precision casting processes as a binder.

Literature Chapter 1

[1.1] Roll, F., Handbuch der Gießerei-Technik, Band I, 1. Teil, Springer Verlag Berlin/Göttingen/Heidelberg, 1959
[1.2] Winnacker-Weingartner, Chemische Technologie Band II, S. 311, Hanser-Verlag München 1950
[1.3] Hinz, W., Silikate, Verlag für Bauwesen Berlin, 1963
[1.4] Petrzela, L., CSR-Patent Nr. 81931 Wasserglas-CO_2-Verfahren, angemeldet 12.12.1947
[1.5] Petrzela, L., Freiberger Forschungsheft B 11, 1956
[1.6] Ljass, A. M., Litejnoe proizvodstvo in Deutsch, 1961
[1.7] Ljass, A. M., Vortrag 28. Internationaler Gießereikongress Wien 1961
[1.8] Gettwert, G., Richarz, F., Neue Ergebnisse über das Kohlensäure-Erstarrungsverfahren, GIESSEREI 59, 1972, Nr. 22, S. 649–654
[1.9] Gerstmann, O., Hertel, R. Seidemann, R., GISAZEM – ein umweltfreundliches, schnell selbsthärtendes Bindersystem, Gießereitechnik 23, 1977, Nr. 4, S. 101–103
[1.10] Nishyama, T., Nach dem Nishiyama-Verfahren hergestellte exothermisch selbsthärtende Formen, GIESSEREI 51, 1964, Nr. 7, S. 167–172
[1.11] Klose, G. R., Fließfähige selbsthärtende Formstoffe, GIESSEREI 59, 1972, Nr. 5, S. 139–146
[1.12] Anwenderinformationen Gisacodur-Verfahren, Leipzig, 1978
[1.13] MacDonald, R. M., Foundry World, 1979, Nr. 1

2 The Development and State of the Application up to the Year 2000

The period around the turn of the millennium marked, more or less, the lowest point in the application of chemically curing inorganic material moulding systems. Because priorities of productivity, property level, and efficiency rapidly came into the foreground, the foundries began to focus primarily on the application of organic binder systems for mould and core production. The main differences here were between the iron and steel foundries, and the nonferrous foundries.

While iron and most of the steel foundries, at least in the serial casting area, relied almost exclusively on the familiar organic binders, (the main reason being better productivity, higher process reliability, and better mechanical properties) from around the year 2000 there were a number of aluminium foundries and copper foundries that had worked either partially or completely with inorganic binder systems.

The reasons for this are many. Considering first of all that today one still can find many light alloy foundries using primarily or exclusively water glass bonded cores with carbon dioxide gassing in their core production while employing the water glass CO_2 process. The materials used are not dangerous and the work area is not saddled with the cost-intensive disposal procedures for sand and other waste material. The complexity of the produced cores range from simple geometries, such as drill core, up to moderate core geometries (See figures 2.1 and 2.2). Although reliable figures are not available, at this time one can surmise a procedure portion of 5–7 % of total core production for the water glass CO_2 process.

Figure 2.1: Typical water glass CO_2 cores for casting fittings

In 2000 there were two main inorganic processes of cold self-curing mouldmaking. The water glass ester process was used in aluminium foundries for moulding material and core production in the area of hand shape casting up to casting masses of about 1 t. Rea-

Figure 2.2: Typical water glass CO_2 core with steel insert

Figure 2.3: Water glass ester mould from the area of the aluminum casting (photo: Pinter Guss GmbH, Deggendorf)

sons for this were, and still are, the favorable workplace and environment conditions combined with the relatively unproblematic behaviour in the aluminium casting process (fig. 2.3). The second process of self-curing moulding process is the cement moulding process. Although these are used in Germany at the moment in a pronounced niche for the manufacture of marine propellers, it is in this exotic application of the moulding material process wherein lies the appeal of this 'dinosaur' of moulding material process. For the aluminium bronze propellers with maximum mass of about 130 t (i.e. about 160 t of molten metal) required moulding materials must remain absolutely thermally and mechanically stable over a period of up to 20 hours to produce a geometrically exact propeller. For this reason this is the only suitable process to this day. Because of good work place and environmental characteristics in addition to the recycling options of the resulting used sand, this process has the potential to be used increasingly more in the future (See fig. 2.4 and 2.5).

In addition to other exotic inorganic binder systems such as gypsum, silica sol should be mentioned when discussing the whole group of inorganic binder structures. This

Figure 2.4: Cement bonded moulds for ship propellers (photo: MMG Waren GmbH)

Figure 2.5: Aluminum bronze propeller (photo: MMG Waren GmbH)

binder system, which is based on aqueous silica, is one of two which can be used for the preparation of investment casting shell moulding materials in useable binders. Chemically related to water glass binders but in contrast, only containing small amounts of Na_2O, it represents the most important binder system used in the lost wax casting process. The advantage of aqueous systems, in contrast to alcoholic binders (ethyl silicate), is their behaviour in the workplace and environment even though there are increased drying costs in the production of shell moulding materials.

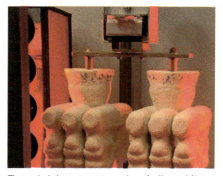

Figure 2.6: Investment casting shell moulding materials (photo: W. Weihnacht)

If one disregards the binder systems for the investment casting process, and one considers the area of casting production with lost moulds (and cores), i.e., using the sand casting process, it can be assumed that around the turn of the millennium 5–10 % of all moulds were produced with inorganic binder systems.

3 Overview of Currently Available Inorganic Binder Systems

3.1 Alkali silicate binder (water glass binder)

The use of aqueous alkali silicate solutions, in particular sodium and in exceptional cases also potassium silicate solutions (better known as sodium or potassium water glass solutions), as a binder for foundry moulding materials dates back to a patent from *Petrzela* in 1947, as already mentioned at the beginning of this book. The adhesive effect of these systems was already known at the end of the 19th century in England and was also described in a corresponding patent but found no practical application. The common term of water glass comes originally from the chemist *Johann Nepomuk von Fuchs* [3.1].

Water glass is not a single chemical compound, but a collective name for glassy solidified melts of alkali silicates of varying composition, and for their solutions. Water glasses can therefore be described as alkaline salts of silicic acids. Water glass solutions can be characterized by their volume or molar ratios of silicic acid (SiO_2) and alkaline oxide (Me_2O), or the weight (modulus), as well as density and viscosity. In addition to the density in g/cm³, the indication in degrees Baume (°Be) in Germany and

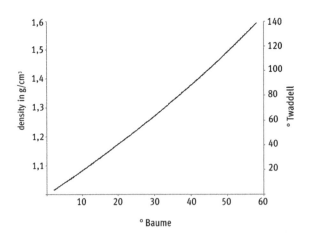

Figure 3.1: Correlations between the density values for water glass solutions in g/cm³, degree Baume (°Be) and Twaddell degrees (°Tw) [3.2]

Twadell in degrees (°Tw) in the UK is widespread. To convert the density units there are the following co-relationships: °Be = 144.3 − (144.3 : density) and Tw = 200° x (density-1). Figure 3.1 shows these co-relationships.

The production of commercial alkali silicates is represented by *Gettwert* in figure 3.2. The most important process today is the fusion. By this method, sodium and potassium silicates can be produced. These serve as raw materials of high purity silica sand, alkali carbonate (soda for sodium, potash for potassium silicate), or alkali as alkali components. The purity of the materials is very important as impurities affect the quality of the alkali silicate solutions generated. In the fusion method, the mixture is then melted in continuously working furnaces at temperatures between 1300 °C and 1500 °C to alkali silicate. The melt is then cooled down abruptly (on rotating steel rolls for example). The resultant piece of glass is then dissolved at temperatures between 140 °C and 180 °C, and at a pressure of 4–9 bar in the autoclave. In this step, the subsequent modulus of the binder solution is adjusted by the addition of water.

The sintering process and the hydrothermal method are applicable only for certain alkali-silicate compositions. The sintering process is used for the production of anhy-

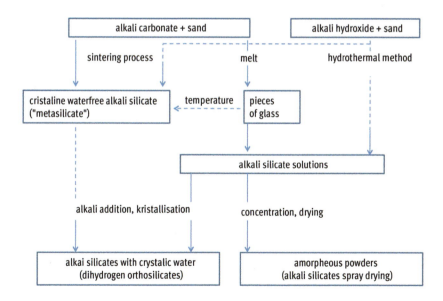

Figure 3.2: Flow diagram for the production of commercial alkali silicates [3.2]

drous sodium metasilicate, which for foundry technology has no purpose. The hydrothermal method is applied to the second binder manufacturing production process. Silica sand and sodium hydroxide are thereby placed in an autoclave under elevated temperature and pressure to reach reaction. This procedure is more energy favorable (compared to the melt method) but it also has, for example, a lower rate of dissolution which increases the cost. In some literature it is sometimes reported that these two processes produce different properties in liquid glass, but other sources refute this claim. Sodium and potassium water glass can be produced using the melting, sintering and the hydrothermal process. Lithium silicates cannot be produced via melting process.

The modulus of the alkali silicate solutions

In his writing *Gettwert* also detailed the properties of water glass solutions (3.2). The general composition of an alkali metal silicate or water glass solution may be represented as follows:

$xSiO_2 * yM_2O * zH_2O$,

where the principle M stands for the alkali metals sodium Na, potassium K or lithium, Li.

A characteristic feature of the alkali silicate solutions is the ratio of silica to alkali metal oxide, which is represented as the modulus. One should make a distinction between weight and molar ratio. The correlations between weight ratio (LVN and molar ratio (MR) with sodium, potassium and lithium silicates are shown in table 3.1.

From the historical development a distinction is made between so-called
- alkaline liquid glasses with MR 2.5,
- neutral liquid glasses with 2.5 – 3.4 MR,
- and (high silica) acids water glasses with MR 3.4.

Table 3.1: Correlation between weight ratio (LVN) and molar ratio (MR).

Lithium silicate	MR = LVN x 0.497
Sodium silicate	MR = LVN x 1.032
Potassium silicate	MR = LVN x 1,566

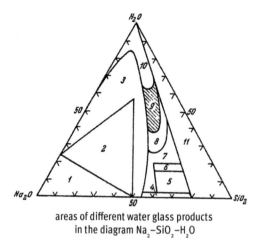

areas of different water glass products
in the diagram $Na_2-SiO_2-H_2O$

Figure 3.3: Ternary system Na_2O-SiO_2-H_2O with the position applied in the foundry silicate solutions [3.3]

In the case of sodium water glass (which is mainly used), the ratio of SiO_2 to N_2O is in most cases 2.0 to 3.3. This modulus has an impact on specific technological properties of the binder systems and moulds. In his detailed post [3.3] *Böhmer* already notes that next to other characteristics of the water glass system that those with a lower modulus generate higher strengths. He also further noted that for conclusions about the properties of water glass films, it is necessary to observe that the material strength is composed of the two components. This is, on the one hand, the adhesion of the binder to the sand grain (complete saturation corresponds to good adhesion), and on the other hand, the strength of the binder bridge itself (i.e. cohesion). *Hinz* deals in [3.4] and [3.5] inter alia with the finely divided silicates, the colloids. The range of silicate solutions useful for the application as moulding material binders in the foundry in question is shown in figure 3.3 on the basis of the system SiO_2-Na_2O-H_2O. Current compositions of these solutions are between 25–35 % SiO_2, 8–15 % Na_2O, and 50–60 % H_2O. In addition to the modulus, silicate binder solutions are characterized also by the parameters of density and viscosity.

The density of alkali silicate solutions

For both sodium solutions and potassium silicate solutions there are several relationships for change of density. At constant concentration and decreasing molar ratio the density of the solution increases. Similarly, increasing densities can be seen with increasing concentration and constant modulus. As with many other chemical com-

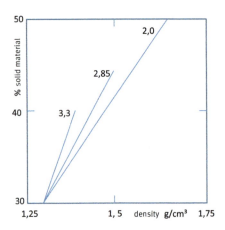

Figure 3.4: Density of sodium silicate solutions with different moduli as a function of solids [3.2]

pounds, the density decreases with increasing temperature of the solution. The density of a sodium silicate solution therefore depends mainly on the modulus ($SiO_2 : Na_2O$). When this increases the density decreases. This is due to the decreasing levels of alkali ions. Since some of the alkali ions in the solution are able to penetrate into the cavities of the silicic acid micelle, increasing the solids content (and thus density) without an increase in volume, this increases the density with inversely decreasing modulus. Influence on density is shown in figure 3.4. The lower the modulus is, the higher the density at the same solids content. The solids content of commercially available alkali silicate solution is between 30 % and 50 % with densities between 1.25 g/cm³ and 1.6 g/cm³. The density exerts an influence on the processing properties and increasing density is to be expected with complicated mixing processes. However, since most of today's water glass binders are at density of 1.5 g/cm³, this has little practical significance.

The viscosity of alkali silicate solutions

One of the most important properties of material moulding binder systems is their viscosity. Viscosity is a measure of a fluid's resistance to flow. The reciprocal of viscosity is fluidity which is a measure of the flow of a fluid. The greater the viscosity, the more viscous (less flowable) the fluid. The lower the viscosity the less viscous (more flowable) the fluid is. Therefore viscosity exerts decisive influence on the mould mixture created with the binder system. The goal would be to use a binder with as low as possible viscosity because it improves the fluidity of the moulding materials. The viscosity of sodium silicate solutions is influenced by several factors. The most important are modulus, concentration of the solution (solids content), temperature, and impurities. The extent to which the solids content affects the viscosity of sodium silicate solutions from 0.0 to 4.0 modulus are shown in figure 3.5 to 3.7.

As the pictures show, the viscosities of solutions with higher modulus rise much more than the solutions with lower modulus. In very dilute solutions (2.0 modulus), low con-

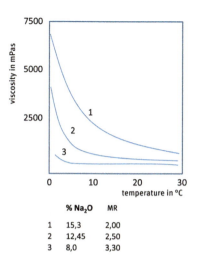

Figure 3.5: Dependence of the viscosity of water glass solutions from the module and the temperature [3.2]

centration increases and only minor viscosity increases occur. When only minor increases in modulus occur (3.4 modulus), this results in immediate increases in viscosity. This means that silicate solutions with high modulus thicken and gel rapidly during concentration or drying. According to figure 3.5, the viscosity decreases generally with increasing temperature. Solutions with a lower molar ratio are therefore more susceptible to the effects of temperature, and solutions with higher modulus are less so. If one looks at the curve 2 for a solution with modulus 2.5 (a frequently applied composition), it is clear why processing such binder solutions should be done at least at a minimum temperature of 10 °C.

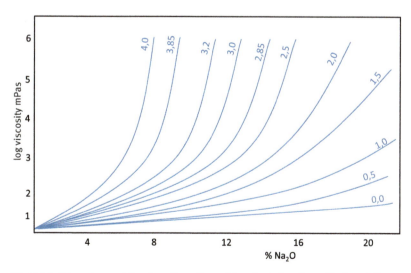

Figure 3.6: Dependency of the viscosity on the solid content of sodium silicate solutions with moduli from 0.0 to 4.0 [3.2]

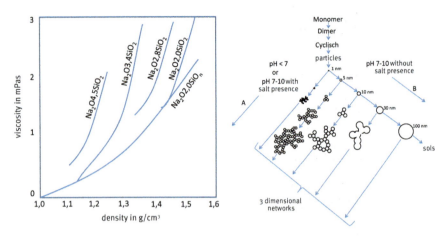

Figure 3.7: Sodium silicate solution: dependence of the viscosity from density and modulus [3.2]

Figure 3.8: Theory for curing by polymerization of water glasses by Iler [3.6]

Polymerization from alkali silicate solutions

In [3.6] Iler presents a comprehensive work where he (among other things) deals with the solidification of water glass. Iler argues that the ions which exist in sodium and potassium silicates are important for foundry applications and explains the production of silicate solutions. The basis for Iler's theory for solidification of water glass is the polymerization of the silicates. Accordingly, the silicate species develop through polymerization (or polycondensation) of the monomeric form through several stages to particles according to their pH. This occurs along with the presence or absence of salts, either three-dimensional gel structure (thus curing the binder films) or continuously growing and forming sol particles. In figure 3.8, this mechanism is shown schematically. In the polymerization the aim is the formation of maximum O-Si-O bonds, siloxane groups, and minimal amounts of uncondensed SiOH silanol groups.

Flemming and Tilch both had a grasp of the structure of water glass solutions together [3.7] and this is schematically shown in figure 3.9. Excluding alkali anions, alkali solutions therefore contain a range of various silicate ions as seminal characteristics. In addition to monomeric ions (such as $[SiO_3]_2$ or $[Si_2O_5]^{2-}$) with increasing concentration of SiO_2 there are also mixtures of condensed silicate anions with more or less advanced networking. In addition, colloid ions (called silica micelle) may occur particularly in solutions with greater than 2.0 modulus. The structure formation of silicate ions and its

Figure 3.9: Scheme for structural composition of sodium silicate solutions by *Flemming* and *Tilch* [3.7]

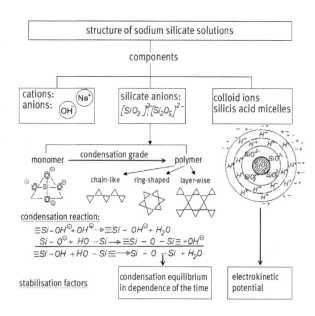

changes happen up to the point of silicate anion balance through condensation or hydrolysis reactions. The equilibrium corresponds to the quantitative distribution of the anion complexes and is dependent on the molar ratio $SiO_2 : M_2O$ and also on the concentration of SiO_2. The stability of the water glass solution compared to a sol-gel conversion is related to the structure of the resulting colloid chemistry and the behaviour of existing silica micelles in preventing coagulation.

These micelles consist of a granule $(SiO_2)_m * (SiO_3)^{2-}{}_x$ and an adsorption layer of H^+ ions. The negative charge of the particles prevents coagulation and gives specific colloid-chemical properties in conjunction with the particle size and the strength of the solvation shell of the solution. An electrokinetic potential difference, the so-called Zeta-potential, is created between the outer layer of the moving particle and the surrounding dispersant. The higher the potential difference is, the more stable the colloid system. Alkaline additives or OH-ions increase the stability of the solution. With addition of electrolytes or acids, the potential difference is canceled: The coagulation of the particles begins when the isoelectric point is achieved. Larger particle associations form out of the colloid particles, causing the bonding of material grains in material moulding binders. This mechanism is the basis for all curing variants of water glass binders. The achievable maximum strengths in solidification by drying are consistent with the previously cited authors.

Jelinek also deals with formation of the silicate binder film during curing in [3.8] and comes to the conclusion that the coagulation will happen at many points simultaneously in moulding materials. The resulting aggregates grow and ultimately join together into the desired three dimensional silicate network. *Jelinek's* further investigations go into this direction in [3.9]. The ^{29}Si NMR spectroscopy represents one way to explain this structure of colloid solutions. On analysis of the chemical shift in the ^{29}Si NMR spectrum, it is possible to make statements about the distribution of the individual silicate anion species with different degrees of cross-linking, i.e. from mono silicate to networked groups (figure 3.10). As shown in figure 3.11, this provides the silicate tetrahedron many structural variants, for example, the group Q3 can occur in 20 varia-

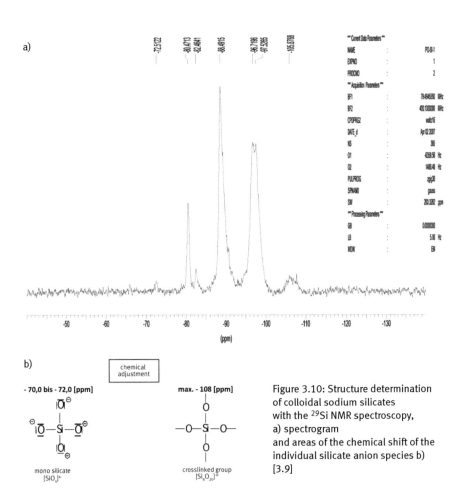

Figure 3.10: Structure determination of colloidal sodium silicates with the ^{29}Si NMR spectroscopy, a) spectrogram and areas of the chemical shift of the individual silicate anion species b) [3.9]

22 | Overview Binder Systems

group description	water glass		
	M = 1,6 Ph = 13,8 KS=12,7% Na$_2$O	M = 2,4 Ph = 13,2 KS=4,2% Na$_2$O	M = 3,24 Ph = 12,10 KS=2,3% Na$_2$O
Q^0	6,8%	1,7%	< 1,0%
Q^1	24,6%	7,3%	4,6%
Q^2	11,5%	2,3%	< 1,0%
Q^2 / Q^3	41,5%	37,5%	27,0%
Q^3	15,6%	45,2%	55,8%
Q^4	0%	6,0%	12,5%

possible structural variations
Q^1 → 4
Q^2 → 10
Q^3 → 20
Q^4 → 35

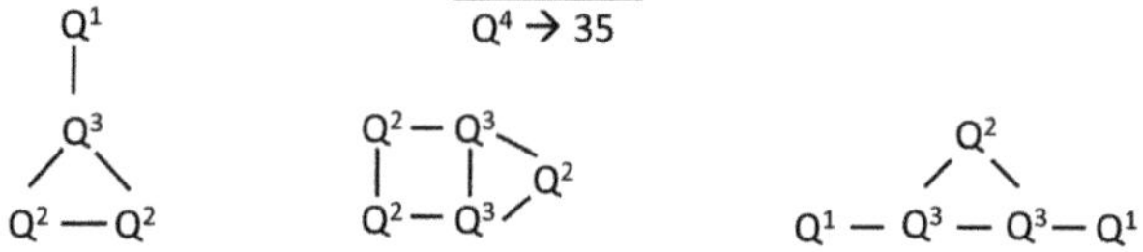

Q^1
 |
 Q^3
 / \
Q^2 — Q^2

Q^2 — Q^3
 | | Q^2
Q^2 — Q^3 /

  Q^2
 / \
Q^1 — Q^3 — Q^3 — Q^1

Figure 3.11: Examples of distributions arising from different silicate anion species and their structural variations in water glass solutions [3.9]

tions. A signal (peak) in a NMR spectrum, therefore, relates to the state of equilibrium of the different anions in the various grades of polycondensation. With the increase of the silicate solution modulus, there is a gradual shift in the amount of mono silicates via cyclic and linear networked forms towards highly cross-linked and branched polysilicate colloid.

The solidification of water glass bonded moulding materials

The deployment and uses of water glass or silicate solutions throughout the industry are so varied that a whole range of solidification options exist in the use of these systems in the foundry. Figure 3.12 shows an overview. Fundamentally there are two basic variants to differentiate: physical solidification (dehydration, drying), and chemical bonding ("chemical curing"). While for decades it was known that very high strengths in water glass cores and moulds are obtained by drying, the early practical application in foundry operation was limited to the chemical bonding models using carbon dioxide or liquid ester hardener. These two processes are still currently applied to solidification technologies. The other possibilities of chemical curing shown in figure 3.12 are no

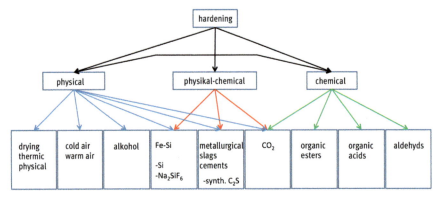

Figure 3.12: Solidification variants of water glass materials

longer used. The only variant that might be interesting again for future developments is the water glass cement method. This cold self-curing method combines the high dimensional stability of the cement mould material with the accelerated (compared to classical cement moulding processes) curing applications of water glass as "hardener" in this system.

On the other hand there are the physical solidification principles that exist in the different possible ways of implementing water glass bonded dry form mixtures. Initially working with heated air, it was found that a tempered core box such as the hot-box method could be used for the manufacture of water glass bonded cores. This is also possible to achieve in a drying oven although this method is time intensive. An acceleration of the solidification can be achieved also with microwave drying. There have been a number of studies done on this in the past (see also figure 3.13). Fig 3.14 shows through the example of a constant binder content of 3 %, the possibilities of the achievable strengths with water glass bonded moulding materials, and the use of different bonding technologies. Knowledge of the differences between physical and chemical bonding was also used in practice several decades ago by the curing of water glass solutions having a modulus of less than or greater than 2.5. Figure 3.15 shows how important technological properties change depending on this parameter. At the moment we see these relationships in the development and market introduction of advanced silicate binder systems hardened in tempered core boxes. This and the currently applied "classic" chemical bonding technologies to mould materials will be discussed in more detail in relevant sections.

The effect of additives in alkali metal silicate solutions

With the classic water glass moulding methods, a number of disadvantages are well known such as partial low strength, poor flowability in the mechanical production of the core, and insufficient shakeout properties. For these and other reasons, the addition of additives to enhance one or more properties was started early on. Often this is referred to as chemical modification. Therefore, even the changing of the module SiO_2 and Na_2O in the binder solution by varying the SiO_2 content is a chemical modification. The SiO_2 content in the binder solution influences the achievable strength properties. Increasing amounts increase the average degree of condensation in the solution and thus the strength of the mould material mixture. Higher alkali content in the moulding material has an unfavorable effect on the shakeout behaviour and sintering tendency. Also, the difficulty of the reclamation of the old moulding materials is increased, as can be expected since increased alkali content is related to increased occurrences of silicate or glass phases. This type of modification however, was used in the practice of

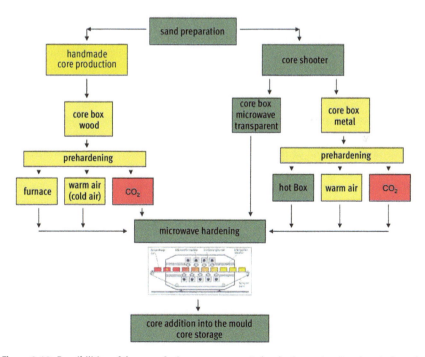

Figure 3.13: Possibilities of the use of microwave energy to hardening water glass bonded sands

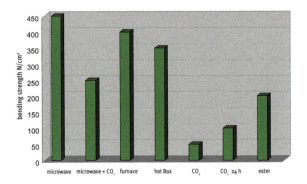

Figure 3.14: Achievable strengths of water glass materials with different hardening technologies (constant binder content of 3 %)

many foundries in the past with water glass binders that differed in modulus and were in use for various applications such as core production. Reference to the procedure is inter alia, also figure 3.15.

The substitution of the sodium ions in the binder solution with other ions also should be referred to as being in the area of chemical modification. In the first place naturally, are the ions of potassium and lithium which play a role in the manufacture and application of the water glass. The introduction of both ions into a sodium silicate solution has specific advantages, but it was rarely practiced due to cost factors. A further modification of this long practiced technology of water glass binders is the addition of organic components such as molasses or dextrin. These small or substantial proportions of binder substances were added to improve, primarily, the decay behaviour through the partial combustion of the binder structures through casting which reduces the residual strength in the aggregate materials. A further effect of such a „sugar modification" can also be an improved flowability of the moulding material during the making of the mechanical core. These binder systems, recognizable by a more or less prounounced brown colouration and frequently a characteristic odour, have the potential for problems through generation of gases. Beginning with a certain total quantity of organic additives they should no longer be considered inorganic binders. The improvement of the properties of water glass binder solutions by inorganic additives is more interesting and therefore has been the subject of a series of research works.

Figure 3.15: Influence of the modulus SiO_2/Na_2O on the technological properties of water-glass-bonded moulding materials [3.7]

sand property	molar ratio			
	MV > 2,5		MV < 2,5	
	hardening technology			
	drying	CO_2 gassing	drying	CO_2 gassing
reactivity	+	+	+	-
bench life	-	-	+	+
storage behaviour	-	-	+	+
strengths	-	-	+	+
abrasion strength	-	-	+	+
residual strength	-	-	+	+

Phosphates as additives in water glass

It has been known for several decades that by adding sodium phosphates to water glass solutions, higher primary strengths are caused in the mould material while the residual strength after thermal stress decreases.

Bialke investigated a whole series of phosphate compounds for their suitability as a modifier [3.10]: Na_3PO_4, $Na_4P_4O_{12}*4\ H_2O$, and $Na_5P_3O_{10}*6H_2O$. The additional amounts of these compounds were between 0.5 and 12.5 % based on the amount of binder. As solidification technologies, the author considers the drying and ester hardening. In regard to resulting strengths (here pressure, splitting, and bending strength), Na_3PO_4 brought the best results. The other phosphates used were sampled in some other areas with good results. In summary, we can state that provided certain peripheral technological conditions are met, the best combination of primary and secondary resistance can be expected with the use of Na_3PO_4. In [3.11] the author of this work expands on the results obtained and refers here to the solidification variants by drying (thermal solidification), CO_2 fumigation, and ester hardening. It is determined the known positive characteristic changes were confirmed by the addition of 3 % Na_3PO_4 whereby the best results were obtained in the thermal solidification. The possibility of the drying of water glass bonded moulding materials has already been addressed but we will still come back to this point later in this book.

Even with *Hähnel* in [3.12], the phosphate modification plays a role. On the basis of molybdate reactive curves it is determined that through certain phosphates (e.g. $Na_3P_4O_9$) there is an increase in the degree of condensation of the silicate anions which can translate into an increased formation of phyllosilicates. Similar statements were also made early on inter alia by *Rönsch* [3.13] whereby in the presence of sodium cyclo-triphosphate the viscosity of a sodium silicate solution rises (changing the degree of condensation) while remaining unchanged upon the addition of sodium tripolyphosphate 6-hydrate. Thus the authors draw a correlation between the degree of condensation and viscosity. This impact on the degree of condensation of the silicate solution can be proven [3.12] with the ^{29}Si and ^{31}P-NMR-Spectroscopy.

In the practical studies in [3.12], the effects of trisodium Na_3PO_4 are examined along with other phosphates to discover that the properties can be controlled by adding amounts of between 0.5 and 2.0 % (figure 3.16 and 3.17). The optimal ratio of water glass solutions is 100:2 with significant increases in strength achieved in reduced residual strengths. From these results, specific applications can be derived for such modified binder systems. For example, as core binder because of higher primary strength, or used in sections of the form that are heavily enclosed in the casting material (because of the lower residual strength and better collabsibility). Since the added phosphate compounds are hazardous to the environment and workplace, the phosphate modification provides a way to improve the properties of water-glass binders. For problem areas in phosphate modification there are a number of other works which build on results presented in previous sections of this publication.

Also dealing with the phosphate modification are publications [3.14] [3.15] [3.16] and [3.17] which deal almost exclusively with the modifier as Na_3PO_4. All these works deal with a number of other questions about water glass moulding processes. Included in these are the considerations of solidification technology by carbon dioxide, water glass cement processing, and liquid ester hardener or ferrosilicon (Nisyama method). Also covered are further questions about reduction of residual strengths, the magnetic influence of binder solutions, and the use of alternative forms of raw materials such as chromite in conjunction with water glass binders. These points will be discussed in the appropriate sections.

**Property influences on water glass binder
solutions through magnetic field treatment**

In [3.18] *Krauße* investigates the subject of magnetic field treatment. By this, strength enhancing effects can be observed. The amount of strength increase is dependent on the SiO_2 content in which improvements are achievable with decreasing of SiO_2. Furthermore, longer exposure time to the magnetic field is also strength increasing. This is explained by the water release related to the action of the magnetic field. In addition, the influence of viscosity and coagulation threshold on the attainable strength has been observed by the author of this manuscript. With the increase of viscosity and coagulation threshold, strength also increases. Here the interval of the coagulation threshold of sodium silicate solutions to achieve strength gains is defined to a range between 0.5 to 1.5 mmol/g.

The extensive work of *Jelinek and others* [3.9] also deals with the magnetic treatment of water glass solutions. It is stated that some foundries used this technology to some success in former Czechoslovak Republic although the ongoing phenomena are still not clearly understood. One explanation is given by *Barabasin* [3.20] in that the treatment of a magnetic field is used to destroy the complex aggregates present in the binder solution through the pulsating exchange field. Accordingly, the viscosity, conductivity, and the

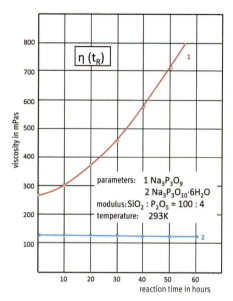

Figure 3.16: Relationship between dynamic viscosity and the duration of action of different phosphates [3.12]

particle size influence the modification effect. In [3.20] the effect is inter alia interpreted through the polar orientation of the molecules. This experiment investigates the effect of a direct current magnetic field on the potential energy, the surface tension, the contact angle and the attainable strength. In this case, alkali silicate solutions with a modulus from 1.2 to 2.8 are examined as the best results were given with those with a modulus of less than 2.0.

With *Cajkova* in [3.21], the free water content of the alkali silicate solution is altered by the magnetic field treatment. With the use of a permanent magnet, a sodium silicate solution with a coagulation threshold between 3.5 and 4.0, as well as induction values of 0.2 to 0.3 t, achieved a maximum change of solvation, i.e. an increase of the free water content. Furthermore, a change in electrokinetic potential (zeta potential) was detected. Other experiments in this study showed that the morphology of the silica gel can be influenced. Binder solutions in the magnetic field thus treated showed a homogeneous and fine-grained structure with higher cohesive strengths. Extensive studies on the topic are found in *Schumann* [3.22], whereby the most influential factors were found to be the field strength of the alternating or permanent field, the direction of the lines of flux to the flow direction of the water glass, and the flow speed.

Influence on the properties of water glass binder solutions by ultrasonic treatment

One of the methods examined in previous studies for improving the properties of the water glass binder solutions is the ultrasonic treatment of the silicate solutions. In one of these studies *Ohmann* deals with this particular method [3.23]. Different types of water glass were subjected to between 10 to 45 minutes of sonication. From the moulding mixtures thus prepared test specimens were produced which were cured either by drying or gassing with carbon dioxide. By identifying the strength values, property improvements and lowered secondary or residual strengths could be determined. However, a clear explanation of the processes involved with ultrasonic treatment and the changes to the binder will not be fully discussed here. Another report on this method of property manipulation of water glass solutions is given in [3.19]. An overview of the most important modification work carried out at the Foundry Institute of the Freiberg Technical University is provided by *Flemming and others* in [3.29] and [3.30].

Synopsis of water glass

Binder systems based on alkali metal silicate or water glass solutions have been used for a long time in the group of chemically curing binder systems that are used at foundries. The variety of solidification options are evidenced through the later considered moulding processes of water glass CO_2, water glass ester, or the water glass hot air method, for example. In the past, these procedures were part of the range of

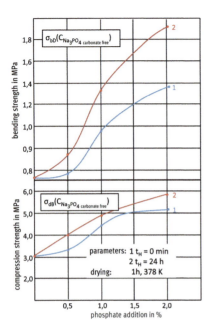

Figure 3.17: Influence of phosphate addition to sodium water glass on bending and compressive strength in the drying of the moulding material mixture [3.12]

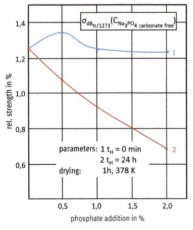

Figure 3.18: Influence of phosphate addition to sodium water glass on the relative residual strength of the moulding material mixture [3.12]

available moulding right up to the present day although with steadily declining proportion as the new millennium approached. In the development of the so called new organic binder systems for the last 10 years the classic water glass binder system played a prominent role.

3.2 Silica Sol for Investment Casting

The silica sols used for the fine casting process are chemically very similar to the alkali silicate or water glass binder solutions described in the previous sections. Since this book deals only with inorganic binders, the other aqueous solutions of amorphous silica in the form of colloidal solutions will be dealt with further in forthcoming books. Silica sols are colloidal silica aqueous solutions, which are prepared by ion exchange from soda water glass. Na^+ ions are replaced with H^+ ions. The amorphous, water-dispersed SiO_2 particles have spherical shape. Their size and SiO_2 concentration are selectable in the manufacturing process. For use as an investment casting binder itself SiO_2 concentrations of 22 % to 30 % and particle size of 8 nm to 15 nm (depending on the application) have been successful. The sols used are Na_2O-stabilized and thus have a very long shelf life. The Na_2O content in investment casting binder systems today are usually below 1 %. Higher levels would greatly impair the sintering behaviour and the fire resistance of the fired investment casting moulds. The hardening of silica sol binders invariably happens by evaporation of the surrounding water and progressive coagulation of SiO_2 particles. The surrounding refractory particles stick together a gel state.

In addition to the SiO_2 content and alkalinity, silica sol solutions are characterized mainly by their viscosity and pH. A pH of between 4 and 8 means increased gel formation which leads to a measurable resistance in the moulded composite material. The initial pH values of the solutions used are approximately between 9.5 and 10.5. The viscosity of the binder solution allows conclusions for its workability. The binding properties are generally based on a gel formation, which is directly related to condensation reaction and agglomeration processes [3.7]. During the mould production the binders are subject to the sol-gel process. This process can be divided into several steps: hydrolysis reaction, condensation reaction and the subsequent drying, and conversion to a ceramic. As previously mentioned, the hydrolysis is used for the preparation of the solutions and the condensation reaction is responsible for the elimination of water, resulting in the forming of a network of silica gel particles of high molecular weight. Silica gel, which can be present in elastic to solid state, can be understood here as colloidal silica. The hydrolysis and condensation are dependent on pH value, this means it works in both bases and acids as a catalyst. Therefore, the hydrolysis is accelerated at high and low pH values.

3.3 Cement as a Binder System

Cement is a finely ground binder cured hydraulically through water separation and includes calcium oxide and silica and alumina and iron oxide as its main components. This binder system which is well known in the construction industry has been used since the 1940s for the production of moulds and cores in the foundry and therefore probably represents the oldest chemical binder. The chemical composition of cement is determined by the so-called "hydraulic modulus", which expresses the mass ratio of basic and acidic components. Practical values for the hydraulic module are 1.7 to 2.2.

$$\text{Hydraulic modulus} = \frac{CaO}{SiO_2 + Al_2O_3 + Fe_2O_3}$$

There are a whole range of cements that are used today for a variety of applications. Table 3.2 summarizes the cement standards and content and shows the basic classifications.

Table 3.2: Overview of the applicable cement standards [3:24.]

Standard	Content	Types of cement	Strength classes
DIN EN 197-1	Standard cement including normal cement to low hydration heat	CEM I to CEM V	32,5 N and R 42,5 N and R 52,5 N and R
DIN EN 197-4	Blast furnace cement with low early strength	CEM III	42,5 L 52,5 L
DIN 1164-10	Special common cement features: - High sulphate resistance - Lower effectiveness alkali content	CEM I bis CEM V	32,5 N und R 42,5 N und R 52,5 N und R
DIN 1164-11	Cement with short solidification		
DIN 1164-12	Cement with an increased proportion of organic constituents		
DIN EN 11216	Special cement with very lower heat of hydration	VLH III VLH IV VLH V	22,5
DIN EN 14647	Alumina cement	CAC	18 (6 h) + 40 (24 h)

Table 3.3: Main types of cements European standard DIN EN 197-1

Designation		Composition	
CEM I	Portland cement	95–100 M.-%	Portland cement clinker
CEM II	Portland Composite cement	65–94 M.-%	Portland cement clinker and small amounts of other components
CEM III	Slag cement	5–64 M.-%	Portland cement clinker and large quantities of slag
CEM IV	Pozzolanic cement	45–89 M.-%	Portland cement clinker and high proportion of pozzolan or fly ash
CEM V	Composite cement	20–64 M.-%	Portland cement clinker and high proportion of blast furnace slag and Pozzolan or fly ash

These five main types of cement are further divided into twenty seven other types of cements according to the quantities of their main components. At the moment only the CEM I Portland cement has significance in later discussions of cement moulding processes.

Besides distinguishing criteria such as cement types and main components, the cements are divided further into individual strength classes. (Minimum compressive strength after 28 days in N/mm^2):
- 22.5
- 32.5
- 42.5
- 52.5

With the exception of strength class 22.5 (which is only for special cements according to EN 11216) the cements are further subdivided according to their initial strength:
- Low initial strength
  L = Low (only for blast furnace cements after DIN EN 197-4)
- Normal initial strength,
  N = Normal
- High initial strength
  R = Rapid.

Overview Binder Systems | 33

Again, in table 3.4 all strength classes are clearly displayed with their initial and standard strength. The standard designation of the cements is done by specifying the class of cement, the norm reference, the symbol of the cement class, the strength class, and the code letter for the corresponding initial strength.

Table 3.4: Cement strength class according to DIN EN 197-1

Strength class	Compressive strength in N/mm²			
	Initial strength		Normal strength	
	2 days	7 days	28 days	
32,5 N	–	≥ 16	≥ 32,5	≤ 52,5
32,5 R	≥ 10,0	–		
42,5 N	≥ 10,0	–	≥ 42,5	≤ 62,5
42,5 R	≥ 20,0	–		
52,5 N	≥ 20,0	–	≥ 52,5	–
52,5 R	≥ 30,0	–		

For example, Portland slag cement with 6–20 % blast furnace slag with a strength class of 42.5 with high initial strength:
→ Portland slag cement to DIN EN 197 – CEM II / AS 42.5 R [3.25]

For the foundry industry the strengths given in table 3.4 make 2 points obvious:
1. The curing times, i.e. the time it takes to cast a mould, are much smaller than the times specified in the applicable DIN standards to reach the standard strength for the construction industry.
2. As a result, the strengths (to be later discussed) in cement moulding process are lower than in other moulding processes.

The highest importance for the binding and curing behaviour of cement bonded sands for foundry applications have the included calcium compositions. Resulting the used raw materials and the manufacturing technologies of the cement his can include the follow main compositions (table 3.5).

Table 3.5: Main components of cements

Portland cement	trikalcium silicate	$3\ CaO \times SiO_2$ (C3S)
Blast furnace slag cement	dicalcium silicate	$2\ CaO \times SiO_2$ (C2S)
Alumina cement	tricalcium aluminate	$3\ CaO \times Al_2O_3$ (C3A)

Cement represents an absolute niche product or process in the casting application at the moment, with Portland cement primarily used as a binder in the cement moulding process. In the past, cement was also used as a curing agent for the water glass binder (water glass cement process). Upon solidification of cement bonded materials, gels as well as crystalline phases can be formed. In a first step, water is attached to the existing oxides and therefore hydrated particles pass into the gel state. In the second step, the decomposition of silicates, oxides and hydroxides develops with the progressive decomposition of the excretions that are produced by the needle-shaped crystalline precipitates. The generally quite slow curing process can be affected by a variation of water (water: cement ratio) or special additives (accelerators). Generally, the work is conducted in the realm of lower hydration. This means that less water than necessary

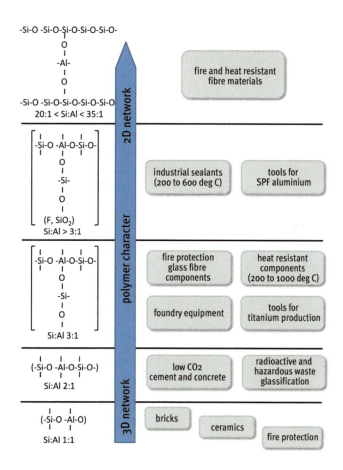

Figure 3.19: Application areas for geopolymers [3.26]

is used for complete curing. Although more addition of water leads to higher ultimate strengths, it also worsens the unpacking and decay behaviour considerably (concrete).

3.4 Geopolymer Binder

According to *Davidovits* [3.26] so-called geo-polymer binders are aluminosilicate solutions that are produced on the basis of water glass. In contrast to the water glass solutions however these systems include not only the composition of tetrahedral Si$(OH)_4$, but also Al$(OH)_4$ tetrahedron. Binder systems with inorganic polymers based on silicon and aluminium (called silicon-oxo-aluminate) have been used in foundries since the 1990s, first in Poland and Czech Republic, and later in Germany. The bond in moulding material comes about through the formation of chains of SiO_4 and AlO_4 tetrahedra. Through the ratio of Si:Al it is possible to change the properties of the system. Figure 3.20 shows an example of the structure of such systems for different ratios of silicon and aluminium. Figure 3.19 shows that with the geopolymer there is a group of substances with a very wide field of application. The importance of the ratio of the two main components is shown in this illustration.

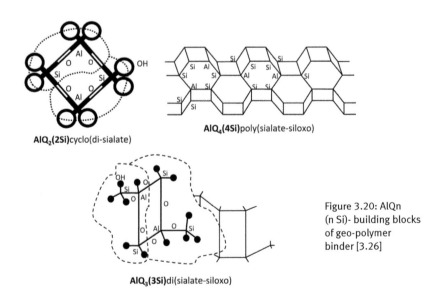

Figure 3.20: AlQn (n Si)- building blocks of geo-polymer binder [3.26]

3.5 Salt Binder Systems

One of the most important innovations in the field of inorganic moulding sand binder in the last 10 years was the development of a group of binder systems which worked on the basis of water-soluble salt compounds. GIFA unveiled this to the public in 2003. This method caused a regular stir in conferences and in journals in the ensuing years. The principle in these systems was to dissolve salts in the water whereby the moistening of the moulding material with the binder was possible. The solidification of the cores produced was through dehydration in a heated core tool („inorganic hot box method"). This way of working presumed from the start that with these methods (large) series cores must be produced to justify the high cost of the heated prototype tools.

This binder systems can be distinguished according to the main component of the binder. In the case of the binder system „Hydro Bond" the main component is sodium polyphosphate and in the binder system „Laempe Kuhs" it is magnesium sulfate. That these binders contained their namesake compounds was communicated from the beginning. Also mentioned was the necessary use of additives. These additives are largely inorganic substances while in some publications smaller amounts of organic additives are also mentioned.

The Hydro Bond binder system is an inorganic water soluble system that is poly phosphate-based (sodium polyphosphate) which is solidified by the dehydration of water-dissolved polyphosphates. The binder consists of salt, water, and special additives. Very few organic ingredients are used which insures unproblematic recycling of core sand waste. The core production is based on conventional core shooting machines. After shooting the cores, the materials are cured by a pressurized hot air gas at temperatures of about 80 °C. This leads to the crystallization of the salt binder and to the strength development of the core.

The company Laempe and Moessner GmbH developed an inorganic binder system in 2001 based on magnesium sulfate, which was patented under the name LK-Binder. First generation LK-Binders are mixtures of pure magnesium sulphate ($MgSO_4 \cdot H_2O$) and mineral substances.

A possible composition according to *Bischoff* [3.27] could be 2–4 wt% anhydrous magnesium sulfate, 0.2 to 0.4 wt% polyphosphate and 3–4 wt% of the total water involved on the basic mould. For the use of tap water, the polyphosphates serve as a preparation liquid carbonation. Thus, the lime contained in tap water can be precipitated *(Sobczyk)* [3:28]. The hydrate of magnesium sulfate can be converted from analog to calcium sulfate in anhydrous or lower levels. Through the absorption of water the solidification of the bonds is achieved and through repeated addition of water, the bonds can be dissolved again.

The pure magnesium sulfate binder system however brought problems in humidity control in that the safety of the process was compromised because of high humidity conditions. Because of this, the main component of $MgSO_4$ has been replaced a mixture of various mineral salts (sulfates, silicates, borates). These binders of the so-called second generation systems are based on silicate („water glass") and are not discussed here further.

An often cited advantage of this group of binder systems was the water solubility of both the binder and the moulding parts produced, and also primarily the water solubility of the cores. The main benefit of this feature is the possibility of wet decoring. At the time this advantage was expressed in a series of speeches and other practical demonstrations during various public events. Early on it became clear that such a technology is practical only in the field of light metal casting, particularly aluminium casting. In the production of iron castings for example, wet decoring would be expected to bring massive corrosion.

The operating temperatures of the core boxes were initially in the vicinity of 200 °C mark but have been corrected over time downwardly towards 150 °C on economic grounds. To reduce the cycle times – which was done in order to compete with the current benchmark of the times in the PUR cold box process – during the early stage of the development, just a preliminary hardening through the edge crust formation in the core moulding tool was employed. The final solidification would be achieved through, for example, rinsing with hot or cold air, or a final microwave curing.

The early stormy development times were hindered repeatedly by setbacks which were largely caused by overly high expectations. For example, efforts to achieve economic regeneration of the sand waste were set too early after attempts through wet regeneration had not brought the desired success. As a result there was a shift early on in the direction of the silicate binder systems which are essentially based on alkali silicate or water glass solutions. For salt binding systems this meant the development activity was basically frozen. This is unfortunate since the potential of these innovative systems is surely not exhausted. There remains the hope that in due course there will be a "revival" of these systems.

Sources and further details about salt binder systems could be given at this point, however, this will be done in the section where the two binder systems and methods are presented in detail.

3.6 Gypsum as Binder

Gypsum ($CaSO_4 * 2H_2O$) is obtained by partial calcination of gypsum. After a firing process and the resultant dehydration stage, different types of gypsum plaster are used. Gypsum can be used in three ways in casting production:
 1. as a binder in conjunction with a moulding material, such as silica sand
 2. as the moulding material in conjunction with water during plaster moulding
 3. as pattern material.

If the plaster is used simultaneously as moulding material and a binder system, we are then talking about the plaster mould process.

Literature – Chapter 3

[3.1] von Fuchs, J. N., Bereitung, Eigenschaften und Nutzanwendung des Wasserglases, Dinglers Polytechnisches Journal, Nr. 17, 1825
[3.2] Gettwert, G., Lösliche Silikate, Firmenschrift zum 100jährigen Firmenjubiläum der Woellner-Werke Ludwigshafen 1996
[3.3] Böhmer, S., Beitrag zur Theorie des Wasserglas-Formverfahrens, Gießereitechnik 1957, Teil 2, Nr. 2, S. 26–32
[3.4] Hinz, W., Silikate; Grundlagen der Silikatwissenschaft und Silikattechnik, Band 1, Verlag für Bauwesen Berlin 1970
[3.5] Hinz, W., wie [3.3] Band 2
[3.6] Iler, R. K., The Chemistry of Silica, Solubility, Polymersation, Colloid and Surface Properties and Biochemistry; Verlag John Wiley & Sons, Inc., New York 1979
[3.7] Flemming, E., Tilch, W., Formstoffe und Formverfahren, Deutscher Verlag für Grundstoffindustrie Stuttgart-Leipzig, 1993
[3.8] Jelinek, P., Aushärtungsprozess von Formstoffmischungen mit alkalischen Silikaten; Konferenzbeitrag 20./21.10.1998 in Zlin, Tschechische Republik
[3.9] Jelinek, P., Polzin, H., Strukturuntersuchungen und Festigkeitseigenschaften von Natrium-Silikat-Bindern, GIESSEREI-PRAXIS, 2003, Nr. 2, S. 51–60
[3.10] Bialke, K., Untersuchungen zum Einfluss von Phosphaten auf die Festigkeitseigenschaften wasserglasgebundener Formstoffe; Studienarbeit TU Bergakademie Freiberg, 1981
[3.11] Bialke, K., Untersuchungen zum Einfluß ausgewählter Modifikatoren auf die verarbeitungs- und gießtechnologischen Eigenschaften wasserglasgebundener Formstoffe, Diplomarbeit, TU Bergakademie Freiberg, 1982

[3.12] Hähnel, U., Beitrag zur Beeinflussung des Festigkeitsverhaltens wasserglasgebundener Formstoffe durch eine physikalische und chemische Modifizierung der Natriumsilikatlösung, Dissertation, TU Bergakademie Freiberg, 1982
[3.13] Rönsch, E., Schneider, W., Scheler, H., Silikattechnik 1975, Nr. 7, S. 232
[3.14] Wolf, U., Erarbeitung und Erprobung einer Kernherstellungstechnologie auf der Basis modifizierter Wasserglaslösungen für Stahlguss; Diplomarbeit, TU Bergakademie Freiberg, 1982
[3.15] Köhler, A., Grundlagenuntersuchungen zur thermischen Härtung von Kernformstoffen auf der Basis modifizierter Wasserglaslösungen; Diplomarbeit, TU Bergakademie Freiberg, 1982
[3.16] Hälsig, F., Einsatz von modifiziertem Wasserglas, Abschlussbericht Ingenieurpraktikum, TU Bergakademie Freiberg, 1983
[3.17] Waßmuth, S., Untersuchungen zu Einsatzmöglichkeiten chemisch modifizierter Wasserglaslösungen in der Kernfertigung, Abschlussbericht Ingenieurpraktikum, TU Bergakademie Freiberg, 1984
[3.18] Krauße, L., Abid, B., Untersuchung von im Magnetfeld behandelten Wasserglaslösungen, Studienarbeit, 1980
[3.19] Jelinek, P., Petrikova, R., Flemming, E., Hähnel, U., Möglichkeiten der Modifizierung von Wasserglaslösungen durch Magnetfeld- und Ultraschallbehandlung und praktische Erfahrungen bei der Anwendung von magnetfeldbehandelten Wasserglaslösungen in der CSSR; Gießereitechnik 1982, Nr. 1, S.21
[3.20] Barabasin, W. S., Litjenoje Proizvodstvo 1971, Nr. 9, S.22
[3.21] Cajkova, R., Freiberger Forschungsheft B 209, S.55
[3.22] Schumann, H., Chemischer Aufbau, Wirkungsweise und Prüfung von Wasserglasbindemitteln, Literaturübersicht, VEB Chemische Werke Cottbus 1978
[3.23] Ohmann, K., Untersuchungen zum Einsatz von US-behandelten Wasserglaslösungen als Formstoffbinder, Studienarbeit, TU Bergakademie Freiberg, 1980
[3.24] Betontechnische Daten, HeidelbergCement AG Heidelberg, 2008, S.2
[3.25] Wrase, M., Untersuchungen zur Optimierung des Zementformverfahren, Großer Beleg, TU Bergakademie Freiberg, 2010
[3.26] Davidovits, J., Geopolymer, Chemistry & Applications, Institut Geopolymere, 1988
[3.27] Bischoff, U., Untersuchungen zum Einsatz eines wasserlöslichen, anorganischen Kernbinders auf Basis von Magnesiumsulfat in einer Aluminium-Leichtmetallgießerei, Dissertation TU BA Freiberg, 2003
[3.28] Sobczyk, M., Untersuchung zur Nutzung der Vakuumtrocknungshärtung für die Herstellung und den Einsatz magnesiumsulfatgebundener Kerne für den Leichtmetallguss, Dissertation Universität Magdeburg, 2008
[3.29] Flemming, E., Mende, H., Beitrag zu Problemen der Qualität und Verwendung von Wasserglaslösungen als Formstoffbinder, Gießereitechnik 35, 1989, Nr. 6, S. 192–197
[3.30] Flemming, E., Schmidt, M., Hähnel, U., Ledwoin, E., Erfahrungen beim Einsatz von modifizierten Wasserglaslösungen, Gießereitechnik 35, 1989, Nr. 10, S. 299–302

4 Classification of Moulding Processes with Inorganic Binder Systems

Chemically curing moulding method can be classified or categorized in different ways. Although this book is limited to inorganic binder systems, generally accepted nomenclature for the chemically cured moulding material systems is discussed in the next sections:

- Solidification by gassing with a gaseous hardener (gas-curing moulding method) at room temperature.
- Solidification at room temperature by internal self-curing (cold self hardening method).
- Curing by tempered prototype tools ("inorganic hot box method").

Advantages and disadvantages of each method on specific priority groups and applications (e.g. mould or core making) will be discussed at the appropriate places. The following sections deal with these inorganic moulding processes according to their classification.

Table 4.1: Classification of inorganic moulding processes

Hardening by gasification moulding process	Cold-self curing moulding process	Warm or hot curing moulding process
Water glass CO_2 process	Cement moulding process	Processes with water glass binders and solidification by drying
Water glass hot-air process	Water glass ester process	AWB® process
	Moulding processes with geopolymer binders	Inotec® process Cordis® process Processes with salt binder systems Investment casting with silica sol binders

Classification | 41

The process variations presented in the column "warm/hot-setting moulding processes" in lines 1 to 4 are primarily warm-box processes with use of silicate or water glass bonding systems, and should therefore really be written on one line. But because these processes are the ones with the currently greatest potential for development and use, they are further differentiated here. In the absence of other names of the binder systems, Cordis® and Inotec® are used at this point.

4.1 Hardening by gasification process

4.1.1 Water glass CO_2 process

The historical importance of the process

The principles of water glass CO_2 process have been known for a long time. The first patent law mentions are found in 1862 by *Gossage* and in 1892 by *Hargreaves and Poulsen*, however at that time an industrial application was not yet realized. The widescale technical application started in 1947 when *Petrzela* introduced another patent with details of a "carbonation solidification" process [4.1]. The development work on the procedure was mainly carried out in the former Wittkowitz ironworks. At this time no other form of chemical curing processes were available. One spoke generally of "chemically hardened moulds and cores" and later of the "CO_2 process". The procedure was soon patented in a number of other European countries such as Italy, Switzerland, France and England. The impetus for the development of the procedure was realized through the lost wax casting method by *Petrzela* through the use of ethyl silicate as a binder system. What was truly revolutionary in the new procedure was that the aqueous silicates (water glass solutions) with the addition of carbon dioxide, cured within minutes!

The water glass-CO_2 process is the forerunner of the moulding processes that use alkali silicate or water glass solutions as binder systems. Its introduction allowed a significant increase in productivity and improvements in properties of produced cores and moulds when compared to the previously applied mainstream mouldings of dry sand, fireclay or cement. Curing was carried out by exposing the mould material mixture with carbon dioxide and thus the first, and still used "cold box gasification method" was born. The basics of this chemical curing are comprehensively covered by *Petrzela* in [4.2].

Therefore, the basis of the chemical hardening in the formation of hydrated silicon dioxide is through the decomposition of aqueous solutions of alkaline silicates and also possibly hydrolysable solutions from the organic ester of the silicon dioxides. The reaction between the sodium silicate and the carbon dioxide can be shown by the following formula:

$$R_2O * nSiO_2 * mH_2O + CO_2 = R_2CO_3 + qH_2O + n(SiO_2 + pH_2O)$$

where R is the alkaline component, that usually corresponds to Na or K.

The state of the SiO_2 hydrogel and therefore also the physical and technological properties of the mould mixtures produced depends on the process of polymerization as the basis of solidification. The carbon dioxide goes into solution in alkaline silicate, and the SiO_2 hydrogel is generated on the contact border of the water glass excess in a neutral medium. Under these conditions the polymerization achieves its greatest speed, because during the process a grainy precipitate forms which has no strength. By slower curing (or drying), massive silicate gels are formed which also have higher strengths. Optimal strengths would therefore be achieved in low molecular silica binder solutions that solidify by gradual polymerization. This however, is not possible in practice and so the foundries have to make compromises in the use of water glass binders. *Petrzela* had already noted that hydrated silica disintegrates well in the temperature range between 400 °C and 500 °C which is important for the application of water glass binder in aluminium casting even today. Conversely, the decay can be worsened by the use of water glass of higher alkalinity. The solidification of water glass bonded moulding material theoretically runs in three stages:

1. In the first stage only a part of the water glass is broken down. The hydrated silicon dioxide which is excreted dissolves in the alkaline water glass residue and the sodium carbonate. In this way semisolid, rubbery binder shells form on the grains of the moulding material.
2. After completion of the neutralization all of the water glass is decomposed.
3. Through the neutralization of the water glass solution the formed sodium carbonate reacts further with the carbon dioxide to form sodium bicarbonate.

According to *Petrzela*, curing begins before the completion of the neutralization. The ultimate strength of the produced moulds and cores will be higher according to the amount of free bases included in the final product. This goal can be achieved by using shorter times of carbon dioxide gassing or by using water glass of low modulus (with high alkalinity). Since using water glass binder with low modulus can be associated with difficulties in processing, using shorter gassing times is more advantageous; this is recommended by *Atterton* [4.3] among others. Furthermore, it is found that highly alkaline (low modulus) water glasses show poor decay behaviour with high residual strength. In order to improve this behaviour, *Petrzela* recommends the addition of organic disintegration aids, a method not currently advisable. In order to keep the gassing with carbon dioxide as short as possible, the use of a carbon dioxide-air mixture is recommended. To quantify the use of the gas curing, *Petrzela* gives following recommendation: 100 kg mould material (with 4–5 % water glass) with 0.3 kg to 0.4 kg carbon dioxide. *Petrzela* conducted the first extensive studies with a water glass modulus 3.5 (36 – 38° Be). He gives an interesting basis for calculating the compressive strength:

compressive strength (kg/cm²) = % water glass -2,5.

The inventor of the process further notes that the bending strengths are lower here than in the production of oil sand cores and can be compared to the strengths achievable when using the dry-casting process. This is sufficient for a whole range of non-complex cores even today. Even so, moisture in the moulding material that is kept unnecessarily long during transport and storage of the cores is to be avoided during post cure. *Petrzela* determined that the problem area for residual strength or decay behaviour after pouring is related to the prevailing post-casting maximum temperatures in the moulding material. Accordingly, the highest strengths develop in areas which have been heated to temperatures between 800 °C and 1000 °C. Lower strengths are developed in the temperature range between 400 °C and 600 °C. This is explained by the formation of alkaline glasses of sodium carbonate and silica in conjunction with increased sintering phenomena from about 800 °C. On further heating there is an onset of increased sintering between the binder system and the mould base material silica sand. At a temperature of 1400 °C an improved decay behaviour is evident, the exact causes of which will not be explored here. However, it is worth noting that the improved decay probably occurs as a result of sintered material stresses and subsequent cracking. This also explains why today more steel foundries work with water glass CO_2 cores than iron foundries. In steel casting the problem of decay behaviour is much less pronounced than in the production of iron castings.

The storage of water glass bonded uncured moulding mixtures requires some special precautions; they must be protected against drying out and also the ingress of carbon dioxide. According to [4.2] the (normal) carbon dioxide in the air is no problem for the workability of the moulding mixture. In this context, there is an interesting example of a steel foundry that processed 25 t of water glass bound material on a saturday. The material was then left open to the atmosphere over the weekend. Only one layer of 10 mm mould material had dried on the surface which was removed on monday. However, it is undoubtedly cheaper to cover the mixed resin to prevent such a drying out and this example is surely no longer conventional for weekend storage of materials. The workability of sodium silicate-bonded moulding substances over a period of a few hours (which is common) is possible when using a sealed container at the workplace or the core shooting machine.

The key to *Petrzela* advantages over the hitherto mainly applied technologies, (e.g. those based on sand moulds which helped in the breakthrough of new methods) should be summarized at this point again briefly:
- Reduction of mould and core production times
- Time, space, and cost savings as a result of no longer needed drying times for moulds and cores
- The possibility of the reclamation of sand waste
- Lower price for materials used
- Good finishes, and reduced cleaning requirements

These benefits can only be realized if certain conditions are met, which is also mentioned in this source. Because this knowledge is unfortunately not available in all foundries, these conditions should be mentioned here again:
- Precise, tightly closing ports with the necessary inputs and outflow provided for the hardening gas core boxes
- Exact observance of technological requirements for the processing of the moulded materials which of course presupposes such requirements exist
- The use of suitable mixing, transportation and storage technology
- reclamation technologies and aggregates suitable for the process (this is currently being subject to various developments)

With the decline of the classic water glass moulding process, knowledge for the reclamation of used sand was lost until reclamation was in fact completely absent from every day production. During the development of the "new" organic method the reclamation of used sand was moved back into the center of attention (this method will also be discussed in this book). Regardless of which path is taken in reclamation (e.g. wet or mechanical) one is certain to find similar practices in the past.

Petrzela also refers to the necessity of gassing with carbon dioxide [4.2]. This step can be applied through hollow needles, rubber bells (probably the most widely used currently), the boring of the core box or pattern (mould production), the core armor or the mould box. Exotic options such as fumigation by the moulding from dry ice or the use of over or under pressurized chambers can also be mentioned here. The pressure range for gassing is given at between 1 to 4 bar with the most frequently used at a middle point 2 bar. To work efficiently, the following guidelines for dealing with carbon dioxide are given:
- The carbon dioxide may be used in the workplace after a reduction of pressure
- All gas components (gas flushing, valves, pipes) must be gastight

- Freezing of the gas (e.g. in the reduction valves) can be prevented by heating the CO_2
- Curing should take place from the inside to the outside
- Series production can be automated by the use of time synchronized valves

The theoretical consumption of carbon dioxide is therefore between 0.3 % and 0.4 % (based on the moulding material quantity) however the allowable amount should be only up to 1 %. Already at that time it was shown in operational measurements that the carbon dioxide consumption is often significantly higher than the mentioned optimum addition amounts (usually 9 %). By such a method mould material costs are too prohibitive and the potential strength of the mould material is not depleted sufficiently through over gassing. This is an all too often neglected point and will be discussed again later in this document. Finally, in table 4.1 moulded plastic formulations that were recommended at that time are summarized where a water glass binder system

Table 4.1: Recommended moulding material formulations for the water glass CO_2 process, according to [4.2]

Use of mould material	Silica sand (AGS 0.2–0.3 mm)	Water glass (M -3.0–3.5)	Reclaimed sand	Additives/comments
Cast steel (mould and core)	100 wt %	5 wt%, cores 6 wt %	–	Use of wet sand
Cast steel (mould and core)	15–25 wt %	5–6 wt % (wet reclaimed)	75–8 wt %	Use of wet sand
Cast steel (mould and core)	60 wt % incl.	5 wt % 8–10 % NaOH	–	40 wt% natural sand (up to 15% clay)
Gray and malleable iron	93–97 wt %	5 wt %	–	3–7 wt % coal dust
Gray and malleable iron	15–25 wt %	5–6 wt %	75–85 wt % (wet reclaimed)	3–7 wt % coal dust
Heavy and light metal alloys	93–97 wt %	5 wt %*	–	Castings of Mg alloy no boric acid, S acknowledge
Backfill sand	0–50 wt %	4 wt%	0–50 wt %	50–100 wt % used sand

* For aluminum castings no higher binder contents should be used due to cracking.

with a density of 36°–38° Be is used, which corresponds to a modulus range between 3.0 to 3.5.

In Germany *Böhmer* published extensive work on the theory of the new method [4.4–4.6]. The process was first applied in 1952 and continued in the following years also on an increasingly international basis. According to *Böhmer* the hardening through CO_2 can be simplified as displacement or precipitate reaction:

$$Na_2SiO_3 + CO_2 = Na_2CO_3 + SiO_2$$

The dissolved and dissociated base materials react forming silicic acid and soda. Silicic acid separates amorphously and is more or less hydrated. The amorphous silicic acid hydrate is finally responsible for the adhesion of the moulding material [4.4]. It is further determined that the strength-bearing amorphous silicic acid precipitate holds the majority of the existing water, while some of the water is removed from the mould material by the carbon dioxide flowing through the mold material debris. With excess gassing one can expect property deterioration. The subject of excess gassing will be returned to at a later point.

Schmidt and *Philip* also deal with the characterization of the hardening process [4.7]. The authors start from the decomposition equation of the water glass and deal in detail with the coagulation of silicic acid. The portrayal of the setting mechanism is based on the hydrolysis of the silicate solution and describes the process of coagulation by the progressive elimination of OH ions surplus. The coagulation proceeds accordingly with elimination of water according to the following scheme:

$$Na_2H_2SiO_4 + 2\ H_2O = H_4SiO_4 + 2\ Na^+ + 2OH^-$$

$$2\ Si(OH)_4 = (OH)_3Si - O - Si(OH)_3 + H_2O$$

The interesting thing about this source is the finding that the amorphous silica caused by complete polymerization has a low binding strength due to its structure and that the intercalated carbonate system takes a significant part of the resistance itself. Around the same time as in the former Czechoslovak Republic (*Petrzela*) A. M. *Ljass* was developing the water glass CO_2 method in the USSR [4.8 and 4.9]. *Ljass* differentiated the curing process in three phases:
1. Decomposition of the sodium silicate.
2. Formation of the silica gel.
3. Solidification of the gel by water discharge.

This results in the following equations:
1. $Na_2Si_2O_5 + CO_2 = Na_2CO_3 + 2\ SiO_2$
2. $m\ SiO_2 + n\ H_2O = m\ SiO_2 \times n\ H_2O$
3. $m\ SiO_2 \times n\ H_2O = SiO_2 \times p\ H_2O(n-p)H_2O$

Therefore, the best strength for a silicic acid hydrate will be at about 13 % water content. The topic of solidification by water dehydratation (drying) will be revisited later in this document.

Concerning the use of water glass solutions in the German foundry industry, *Böhmer* had already noted that primarily viscous solutions with about 50 % water and a density of about 1.5 g/cm³ (50° Be) were used [4.4]. The modulus usually lies well above 2.0; it was recognized early on that it was not a simple precipitation reaction from CO_2 water glass hardening. As a consequence, a series of experiments were conducted in order to better interpret the phenomena involved.

In addition to structural considerations, *Böhmer*, in [4.5] with the various curing options for the water glass binder system. The effect of carbon dioxide (as with other electrolytes particularly acids and acid salts) is based on a modification of the charge state of the colloidal solution, which is reflected in the shift of the pH. The repulsive forces are reduced between the colloid particles, resulting in the subsequent coagulation. According to *Böhmer*, with the addition of even small amounts of carbon dioxide a small jump in viscosity can be expected. By further (rapid) addition of CO_2 the segregation phenomenon is accelerated; complete segregation being related to a relatively low strength. The strength as a function of carbon dioxide absorption therefore has a maximum. For practical application this is of great importance. It means that for the curing of water glass binder systems a much smaller amount of carbon dioxide is necessary than would be stoichiometrically for the formation of soda. With the progressive separation of the gels formed (curing) the water glass gels begin to shrink under the emission of solvent or electrolyte. A whitish efflorescence of the soda can be observed in the presence of sodium ions or anions of carbonic acid. *Böhmer* these interrelatioships very nicely making reference to the thin films of water glass which are monitored for set periods of time (separation).

According to *Zifferer* [4.10] the pH of the highly alkaline colloid solution decreases through the CO_2 gassing. As a result, the micelle in the forming acidic medium loses the stabilizing diffusion layer whereby the electrokinetic potential of the particle decreases. The consequence is the emergence of a large change in free energy [4.11]. This enthalpy value may decrease but only in simultaneous growth of large particles at the expense of smaller components (reduction of the specific surface area). Since the solubility of colloidal substances in a dispersed phase is very small, particle increase as a result of coagulation is promoted. In the coagulation process there is now a gel between the grains of the mould materials (i.e. in most cases the silica grains) that forms a connection and leads to the hardening of the moulding (mould or core). According to *Jelinek* [4.11], the determining factor for this gel is the ion-particle aggregate, a pH-dependent process. Coagulation is fastest in the range of the isoelectric point (pH range between 5.5 and 6.5) and thus in the area of maximum instability of the colloidal system. From this it can be deduced that there is an increasingly stable system with a decreasing SiO_2/Na_2O modulus while the process of aggregation (i.e. solidification) is increasingly slowed. This in turn means that carbon dioxide, necessary for stabilization, increases.

Some authors [i.e. 1.8, 3.7, 4.5, 4.6, 4.7, 4.12 to 4.16] state that parallel to this process of chemical reaction, a reaction of the silica gel drying occurs where bound aggregates and non-elastic solid particles are formed. The truth and technical significance of this could be questioned in light of the very short gassing times ranging from a few seconds up to 1, 2 or 3 minutes. Finally, *Böhmer* discusses the processing properties of water-glass-bonded moulding materials. Thus at the time, the customary use of natural moulding sands for the method was not recommended since often the mineral limonite is present in these sands, which could interfere with the curing of the water glass film. Also, the use of completely dry or excessively moist silica sand is not recommended. While in excessively dry sand the grain surface will attract water from the binder system (which is no longer available for solidification reactions), in excessively wet sand the opposite is the case, i.e. a softening or dissolution of the binder cover can be expected. In addition the sources point to the reaction of generated test bodies of water. Here the strength is reduced in the composite grains which disintegrate quickly. It is interpreted that with carbon dioxide curing no "melting" of the colloidal particles on the main valences occur nor does complete condensation.

The residual strength (i.e. the strength of the cast) plays a central role in *Böhmer's* considerations [4.4 to 4.6]. These high-temperature properties are dependent on various factors such as the strength of the binder bridges, the nature and intensity of solidification, rate of heating, temperature exposure time and lastly, the cooling rate. The first insights into the behaviour of the mould materials upon heating is shown in fig 4.1 to 4.3.

Therefore the moulding material is first softened in the temperature range of between 150°C and 250°C. The temperature of the initial state of the binder, the thickness

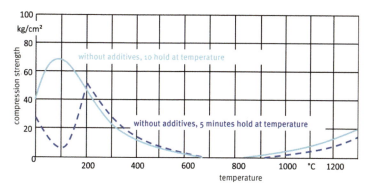

Figure 4.1: Compressive strength after heating and cooling at different holding times [4.3]

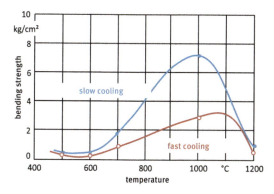

Figure 4.2: The effect of cooling rate on the strength of heated samples [4.4]

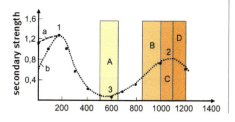

a strong reaction
b small reaction

1 1. maximum secondary strength
2 2. maximum secondary strength
3 minimum secondary strength

area of temperature load
A ligth metal casting
B heavy metal casting
C cast iron
D steel casting

Figure 4.3: The strength of the moulding material after heating to casting temperatures in different alloys [4.4]

Gasification process | 51

of the gel films and the rate of heating are all factors. The previous hardening of the moulded materials is due to the drying and withdrawal of free water. With increasing heating rate and film thickness, this softening temperature moves downward. Already here it is assumed that this softening may fail partially or completely at correspondingly high primary resistance (e.g. due to the drying of the mould material). This could be an explanation for the decay problems that occasionally occur in water glass materials that have been solidified by drying. This softening of the gel film is connected to the evaporation of water, whereby the binder film is blistered and subject to an additional reduction in strength. If this blister formation is reduced and the removal of water is complete, the integrity of the gel is recovered.

With a rise in temperature drying starts in the gel film which leads to an increase of strength in the moulding material. With further increases in temperature, signs of softening in the system are observable and the strength falls off again resulting in an almost complete loss of strength between 500 °C and 600 °C. Depending on their modulus, water glasses melt between 800 °C and 900 °C whereby they completely lose their strength. But if this glassy film solidifies again during subsequent cooling it gives the material form system the high residual strengths by which this method is notorious for. These strengths are influenced by the water in the system. An obvious brittleness occurs with temperature increase (starting at about 1200 °C) resulting in a decline of water content and residual strengths. In contrast, other authors argue that observable sintering between the water and glass mould material occurs from the glass acting as a flux from about 800 °C.

In the 1980s *Svensson* delivered a number of interesting works around the subject of water glass processes with its various solidification options [4.14 and 4.15 among others]. In [4.14] *Svensson* stipulates that the curing of water-glass-bonded moulding materials by gassing with carbon dioxide can best be realized through the use of binder systems with sufficiently high resistance to over-gassing. For the binder solutions with modulus values from 2.0 to 2.4 commonly used in practice, this means that high gasification pressure and flow rates are needed to guarantee the necessary short exposure times. Accordingly, the gas absorption is slower (and thus higher) than in binders with higher modulus. Nevertheless, binder solutions that have the most contact with the carbon dioxide in the area of the mould materials, achieve the maximum strength without being over gassed. All other mould material parts that have not been so strongly fumigated experience strength gains during storage by drying or by casting. In order to use binders with higher modulus, it is possible to reduce the partial pressure of carbon dioxide, without the need to accept loss of mechanical properties. In his work *Svensson* also investigated the influence of additives on sugar-bases (polysorb) and notes that better plastic properties can be achieved with decreasing strength through binder solutions with low modulus. In [4.15] *Svensson* looks closely at the carbon dioxide absorption by silicate binder solution. He notes that low carbon dioxide partial pressures cause the polymerization of small silicate particles which are particularly important for the binder behaviour. In the tests performed it is shown that 1 to 2 mol of carbon dioxide per liter of binder can be dissolved before the formation of sodium carbonate begins. The author holds sodium hydrogen carbonate the most important of the formed carbonates. As noted earlier, the curing can be controlled by the carbon dioxide whereby with increasing modulus the binder solution should be worked at with lower pressures.

The method in more recent times

The following advantages were reasons for the rapid spread of the method according to *Flemming* and Tilch [3.7]:
- In contrast to the time-and energy-consuming dry and oil sand processes, the possibility of curing at room temperature through gassing, and solidification of the cores and moulds occurs in seconds
- This method is universally applicable for single, small or large-series production and easy to handle
- Both the water glass binder and CO_2 hardener can be processed with low environmental impact and are non-flammable, non-offensive in odor and non-toxic

Additional advantages can also be cited:
- The materials used are more cost effective than other binder systems
- Because of the thermoplastic behaviour of water glass bonded cores (and possibly also moulds) this method is suitable for the production of casting alloys that are prone to cracking such as various aluminium or steel casting alloys

In contrast the most important drawbacks are also cited in [3.7]:
- Relatively poor flowability through the binder contained water. Because of the dipole character of water molecules, water-based binders are generally "stickier" than anhydrous systems (e.g. organic systems). The poorer flow behaviour sets boundaries for the production of complex, thin walled cores. Typical applications would be for automotive castings such as cylinder heads
- The strength directly after curing and storage are significantly lower than in organic binder systems
- Limited storage capacity for the produced cores as a result of moisture absorption
- The difficult reclamatation of used sand

The cons listed here are not to be ignored and undoubtedly lead to problems in the manufacturing process of complex castings. The developments over the last few decades however have also provided the above noted progresses. We will come back to these improvements again in this discussion. However, the water glass CO_2 method is still no rival for powerful organic methods such as the polyurethane cold box process. Nevertheless it still has its rationale and probably will have for the foreseeable future if the advantages can still be exploited despite the disadvantages. The water glass CO_2 method is more difficult to handle than organic systems therefore a number of authors suggest that the system binder-hardener should be adjusted as individually as possible on each moulded part [4:16 to 4:18]. Since such a requirement is not practical most binder systems used today are those that can be applied universally in a wide variety of applications.

Flemming and *Tilch* [3.7] indicate that the silica sand used as a moulding material has a crucial importance for expected strength. While you can reach higher strengths for water glass solutions with low modulus (2.0) through the use of fine grained sand, systems with a higher modulus (2.4 or higher) with mean grain sizes of between 0.3 and 0.4 mm are recommended. This relationship is illustrated in Figure 4.4.

The solidification of the moulding mixture is described by *Flemming* and *Tilch* in the following three stages:
- In stage 1, the release of carbon dioxide takes place in the waters of the silicate solution, thereby forming carbonic acid H_2CO_3. Subsequently, the carbon dioxide reacts with the sodium silicate to form unstable disilicic acid $H_2Si_2O_5$.
- In stage 2, this disilicic acid forms a colloidal solution with the water. This leads to the excretion of silica acid micelles and the reduction of the electrokinetic potential.
- In stage 3, a growing neutralization leads to coagulation of particles and the excretion of SiO_2 gels in the shape of grapes, honeycomb or chains.

In addition to these SiO_2 gels as strength bearers, sodium carbonate is formed along with amorphous silicic acid [4.2], [4.5] during excessive gassing. This reduces the strength of the overall system. Also, the formation of sodium bicarbonate may cause a reduction in strength. The amount of the added carbon dioxide in fumigation plays the dominant role for the immediate or initial strengths, i.e. with increased gassing the strength climbs to a maximum. If gassing is continued beyond this maximum, strength decreases due to the destruction of the previously formed gel structure. The reason for this is the developing sodium carbonate needles which enter into the binder bridges and through their unfavorable geometry notching effect, reduce strength.

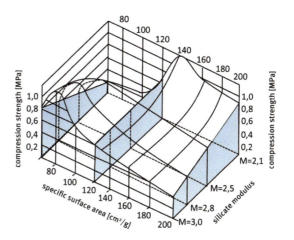

Figure 4.4. The influence of silica sand grain on the strength by the water glass CO_2 process [3.7]

At this point it should be mentioned that there are a number of additional publications that deal with the theory of hardening water glass materials [4.19 to 4.22]. They also come to the conclusion that the gassing with carbon dioxide is based on the chemical reaction of the sodium silicate solution with the formation of silica gels and sodium carbonate and thus the idea that this hardening option leads to the formation of disilicate, sodium carbonate, and strength reducing bicarbonate.

If inadequate carbon dioxide is supplied to the moulding material during fumigation (under-gassing) the initial strength decreases. However, during the subsequent storage times higher strength is gained through drying and formation of glassy sodium silicate. Ideally, during the production of the moulded parts, the gassing times with carbon dioxide should be kept as short as possible in order to exploit the potential strength increases during storage and drying. This means that the produced cores are

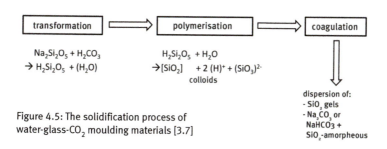

Figure 4.5: The solidification process of water-glass-CO_2 moulding materials [3.7]

Gasification process | 55

generally used after 24 hours. The strength in the moulding material could then be 2 or 3 times that of the immediate strength.

The connection between the modulus and the strength of the silicate solution with CO_2 gassing and drying was shown with figure 3.15 in the previous chapter. As solutions are stable with low modulus, the isoelectric point is reached only after a longer gassing than would be normal for solutions with higher modulus. Better formation and strength increases result from slower curing. Solutions with higher modulus develop their strength much faster and have a comparatively low ultimate strength (Figure 4.8). From these considerations it can be deduced that one should choose a water glass binder that has good processability (e.g. low viscosity and rapid strength gain). Since these two properties alone act in opposing directions, this results in water glass solutions used today that usually have a modulus from 2.3 to 2.5. These binders reach the best balance between the opposed properties. When all of these various points are considered, we see that higher strengths are achieved by drying compared to gassing (see figure 3.15). Therefore, the method of carbon dioxide gassing should only be applied for as long as it takes to achieve sufficient (minimal) initial strength for the handling of the produced cores. The higher the strength through drying, the lower the pre-solidification progression through carbon dioxide. One can observe this in the fact that in most foundries the cores are in storage for at least 24 hours before they are used. To achieve the goal of these higher strengths, the following process variants with low modification and investment measures are implemented:
– Heating of the carbon dioxide
– Gassing with mixtures of air and carbon dioxide which may also be heated
– Use of cold air without CO_2, though this leads to very long exposure times
– Gassing with hot air in the heated core box [4.23] (we will return to this variant on the section dealing with thermo-curing inorganic binder systems).

Further findings are made in [3.7]:
– The storage of manufactured moulded parts in a humid environment (atmosphere in the foundry, soaking in a wet mould) should be avoided as it leads to absorption of water by the hygroscopic water glass cores and results in loss of strength. The critical humidity is 60 % [4.17].
– The amount of gas needed is theoretically between 0.06 and 1 wt %. In practice 1–5 wt % is used when gassing by hand and 1.0 to 1.5 wt % when using automated gassing.
– Applicable gas pressures specified are from 0.05 to 0.15 MPa. At higher pressures there is a risk of leakage of the CO_2 through the core box prior to solidification of the mould material.

Figure 4.6: The effect of gassing and post-curing on the compressive strength of water-glass-bonded sands [3.7]

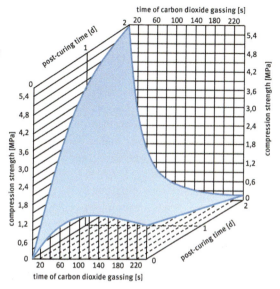

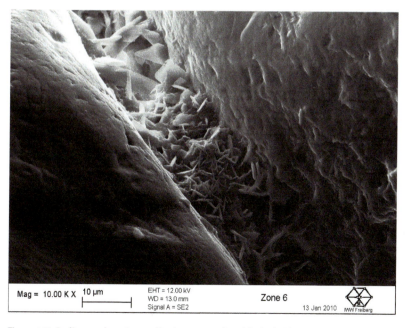

Figure 4.7: Sodium carbonate needles in a water glass binder bridge

- The moulds (i.e. core boxes and patterns) can be made of wood, plastic, metal or composites, which is still a very big advantage of the method.
- During coating application alcohol coating is helpful. If the environmentally friendly water coating is used a drying of the coated cores is necessary.

In 1977, the proportion of water glass CO_2 process used in core production was yet to reach above 14 % [4.24]. In 1986, 7 % was the figure quoted [4.25]. It is additionally noted that in the group of companies with an output of 500 to 1000 tons of casting productions per month, the water glass process was applied 6.9 % of the time with the sharpest decline happening between 1977 and 1986. Today however, it still holds a respectable place in production at 11.6 %. In 1977 there were some companies in this group that still had 34 % of cores produced by the water glass process. The use of the procedure is less than 1 % in the manufacture of moulds [4.26]. According to [4.27] the proportion of cores produced by the water glass CO_2 process in 1990 was 7 %. One can assume that it was after 1986 when this percentage stabilized. Although current figures are not available this percentage seems to be also valid today.

Despite the relatively small optimization potential with use of the water glass CO_2 process (for reasons stated) it is still an indispensable ingredient in the range of moulding process that deal particularly with the production of cores. The foundational reasons are the simple handling and workability of the materials along with economic attractiveness of a binder method needing only simple machinery and technology. The procedure is simply too occupational and environmentally friendly to be dismissed completely. The water glass CO_2 process is still in the production process of a number of small and medium size foundries that deal mainly in the field of aluminium and copper castings. Foundries for iron castings still seek to produce as large a proportion as possible of cores with alkali silicate binders due to the above mentioned advantages. Only for cores with a higher degree of difficulty (i.e. high strength requirements) would one fall back on procedures with a higher property level such as pure cold-box methods.

Even though it is one of the oldest methods, it is therefore still important to be continuously developing improvements in the cold-box method in order to maintain existing production units of this extremely simple organic moulding process. Furthermore, it is important to make available the partially lost knowledge of the CO_2 cured water glass cores for the sake of those who are using this method. For example, it is important to know that the economical use of carbon dioxide in gassing requires that the core must be handled carefully due to its low initial strength but that it also achieves significant strength increase during subsequent storage.

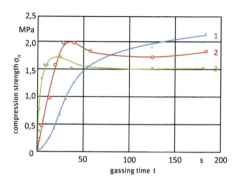

Figure 4.8: Influence of the water glass modulus on the strength development at carbon dioxide gassing [4.17]
1 – modulus 2.0, 4% binder, 2 – modulus 2.5, 4 % binder, 3 – modulus 3.0, 6 % binder

Water glass binder systems today differ significantly from those used 10 years ago in core production. To improve disadvantages such as strength and decay behaviour in the past, various organic additives were used such as molasses and various sugars. These organic components bring proven benefits to the targeted properties, but of course they also change the chemical nature of the binder system. For example if 20 or 25 % of such additives are added to the water glass, it is questionable whether it can still be spoken of as an inorganic binder. Readily observable on such a modified binder is the distinct odor and brownish color.

In [4.28] *Flemming* et al continue to deal with the water glass CO_2 process which in the mid-1990s was virtually dead. In comparing organically modified and unmodified water glass solutions the authors state that increases in strength can be obtained through modification (Fig 4.9] and that with the increasing gassing times the strengths supersede the maximum. Reasons for strength level drops by increased gassing have already been discussed in this document. The gassing pressure in these studies was at 2 bar, the strengths represent immediate strength. The repeatedly cited influence of core storage times on expected strength is also shown in fig 4.10]. This image shows that while the unmodified water glass suffers strength loss after one day (and then stabilizes), two of the three modified binders show continued strength increases up to the end of the measuring period of 7 days. Fig 4.11 also shows decay behaviour and residual strength. The experiments in this case were carried out so that the cylindrical, standard test pieces were exposed to 20 minutes of the temperature indicated in the chart and then after cooling to room temperature were checked on their compressive strength. Image 4.11 shows very nicely that the unmodified water glass with modulus 2 traces the classical residual strength curve for sodium silicate material (see also figure 4.3), while those with module 2.5 fail significantly lower in the maximum. In the three tested modified water glass binders, either the first or the second maximum can be almost suppressed. This means that it is possible to develop water-glass binder systems which are more suitable for use in the iron or aluminium alloy production because of these two suppressed peaks.

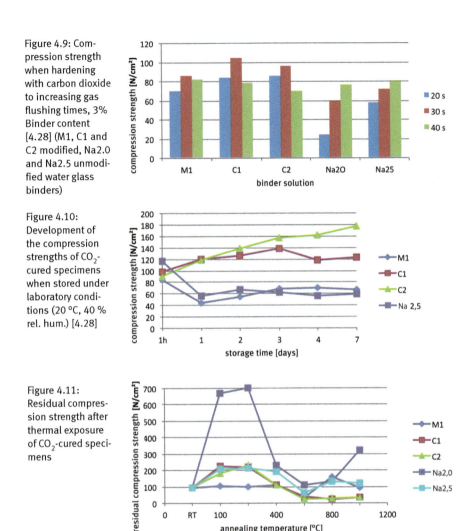

Figure 4.9: Compression strength when hardening with carbon dioxide to increasing gas flushing times, 3% Binder content [4.28] (M1, C1 and C2 modified, Na2.0 and Na2.5 unmodified water glass binders)

Figure 4.10: Development of the compression strengths of CO_2-cured specimens when stored under laboratory conditions (20 °C, 40 % rel. hum.) [4.28]

Figure 4.11: Residual compression strength after thermal exposure of CO_2-cured specimens

In order to reach the desired improvements, only inorganic or possibly very small amounts of inorganic components are allowed today. To attend to the very important issue of moulding mixture fluidity in automated manufacturing of cores using core shooting machines, low viscosity binder solutions are needed; a comparison of flowability using an edge sample according to *Boenisch* [4.29] is shown in fig 4.12. This image shows an example that was shot to 100 % with the binder system B1 and this result is equal to that of a cold box system. Today the practice of taking a sample edge for technological purposes is gone. However the picture does show what impressive results can be achieved in terms of flowability levels available today with water glass systems. The strengths achievable at reasonably short CO_2 gassing and subsequent storage are shown in fig. 4.13. Although the recorded maximum strengths of about 100 N/cm² are significantly lower than the values achievable with other moulding systems, they are still completely sufficient for a wide range of applications [4.30] [4.31].

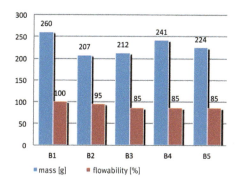

Figure 4.12: flowability of water glass mixtures when using the edge samples [4:33]

4.1.2 Water glass warm-air process

The curing of moulding material on the basis of mixtures of water glass binders can be accomplished by a variety of substances. The two most popular variants of the method with curing by carbon dioxide or liquid ester hardeners are chemical reactions that have been proven to lead to incomplete dehydration of the binder film. Despite the fact that achievable strength of the moulding materials largely depend on the degree of dehydration, maximum values are still reachable. Through the physical consolidation of water glass form substances by the application of drying technology, one can get quite close to the desired maximum water extraction. Strengths obtained by drying the moulded material are about 10 times higher than those from chemical drying meth-

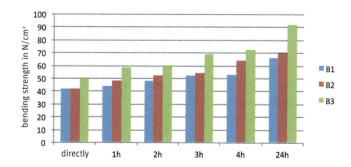

Figure 4.13: Development of bending strength in the storage of water glass CO_2 cores over 24 h, 3 % binder content [4.31]

Gasification process | 61

ods. The drying may be affected by different mechanisms such as drying in still air, cold or hot air, gas flow and drying in hot or warm tools. For all these processes the principle of hardening by dehydration is the same, but the conditions for the transfer of moisture from the system and hence the curing speed, are different.

Gebauer and *Gerstmann* deal in [4.32] with the accelerated solidification using drying chambers, specialized ovens or hot wind. To improve the green strength of the produced parts, the authors propose the addition of bentonite. *Flemming* and *Grisanowitsch* consider drying by heating in [4.33]. They try to model the resulting core materials and consider the behaviour of the formed binder films in dilatometer testing; *Hildebrand* shows in [4.34] that the solidification of water glass mould materials can be realized through warm air gassing.

Tamarante deals with the problem of warm-air gassing in detail in [4.35]. The investigations show that the hot air process achieves high and uniformly distributed strength over the entire cross section of the core. Thin and thick parts are equally cured.

The cores thus produced have an extremely low gas development during casting. The surface quality of the cores is described as very good, which is attributed to the lack of heat cracks in the binder. The subject of thermal curing was also in *Köhler's* work [4.36] in particular drying in ovens at various temperatures. He also puts forth the theory that rapid dehydration to reduced condensation degrees causes lower strength which is contrary to economic aspects. In his studies *Köhler* determined the main areas of water loss during dehydration. Accordingly, 6 % of the water is removed at 100 °C, at 200 °C it is 93 % and at 400 °C it is 100 %. In agreement with other authors *Köhler* recommends curing by a drying binder with a modulus between 2.6 and 3.0 while in carbon dioxide gassing the modulus should be in the range from 2.3 to 2.6. *Köhler* also mentions the possibility of the combined by carbon dioxide and curing with high frequency energy (microwave drying).

According to *Gettwert* [4.37], the viscosity rises sharply in alkali-silicate solutions through concentration (dehydratation) by temperature increase. The water in the solution is bonded in different ways. To completely remove the bound and free water from the solution, you have to use accordingly high temperatures. At 105 °C only the condensation (water or solvent free water) is largely removed. At ca. 200 °C the adsorbed water, meaning water molecules attached to silanol groups (-Si-OH) through hydrogen bonds, is removed. With increasing modulus the drying times decrease and therefore also the residual water content.

Interesting analysis of what takes place during heating of water glass solutions reactions are given in [4.38] by DTA/DTG studies of pure and modified water glass solutions. The DTA curves show endothermic peaks accompanied by weight losses in the DTG gradients. The location of the peaks remained almost unchanged at water glasses with molar ratios from 2.45 to 3.04. However, the mass loss observed in the first peak is larger than the second, and increases with increasing molar ratio of water glass, while those in the second lose weight. The radiographic examinations show that if samples are gradually heated to 1000 °C it results in the forming of crystalline phyllosilicates.

Drying by hot air gassing is also examined by *Svensson* [4.39]. Thereafter, the hot air drying process divides into heat exchange between the hot air and the sand, water evaporation and condensation in the hot air, and the removal of the saturated or supersaturated air from the sand body. The research conducted takes into account the strength development during drying, air temperature, the amount of air, the sodium silicate binder used (binder content) and lastly, sand temperature. As expected the strength increases with decreasing water content and rising temperatures accelerate the drying process. Binder hardening solutions with a modulus of 2.2 have been evaluated to be favorable because the water content to be removed is lower. The sand temperature control is very important in this process because the water solubility is dependent on air temperature.

Carlson and *Thyberg* reported in [4.40] also on the core production by water-glass warm-air process for the manufacturing of cylinder blocks and heads for motor vehicles. Here, a heated core box is used (135 °C to 150 °C) and then curing by hot air (170 °C to 190 °C) is applied. Through this an improved shoot-ability and better disintegration behaviour is realized. In the experience of this author, the combination of hot air and carbon dioxide is beneficial. It is noteworthy in this publication that the procedure was used for mass production.

Penko discusses in detail the curing mechanism in carbon dioxide gassing of water glass binders [4.41]. In his opinion the warm-air gassing or the combination of carbon dioxide and hot air and the manufacturing of cores in 135 °C to 150 °C hot core boxes brings benefits. In [4.42] *Döpp* (among others) states that in the hot air curing (260–300 °C) with 1.5 % water glass, sufficient strengths are already obtained and decay properties are also better. *Jelinek* looks back on the rich tradition of using water glass binder in the Czech Republic [4.43] and puts the colloidal properties of water glass at the beginning of his discussions based on the application of these systems as sand moulding binders. The characteristic features of various hardening possibilities

Figure 4.14: Selection of typical cores water glass CO_2 method images
a) to c) GHM GmbH, images d) to f) MGC GmbH

are discussed including drying by air gassing. Solidification using hot or cold air not only leads to higher strengths, but also improves the storage stability of the produced cores. Warm air drying has the distinct advantage over conventional air-drying in that high strengths through dehydration can be achieved much faster. However, because of the still relatively long duration, increased energy expenditure is necessary.

4.1.3 Warm air drying in water glass powder systems

A special feature in the area of solidification by warm air is the water glass powder system. Spray-dried water glass is used as a binder which in this drying process loses much of its water. The aim of the development was to remove the water (which disturbs the mould material) through the mixing and the solidification processes. The consolidation of the compacted moulding (usually the core) was then carried out by gassing with heated water vapor.

Huusmann and *Lemkov* investigate in [4.44], [4.45] and [4.46] hot air-hardening of the water glass powder with the aim of improving the decay properties which can be essentially achieved by a significant reduction of the binder contents. The experiments were done with binder portions of only 1 %. The specimens were solidified with steam or with air heated to between 130 °C and 150 °C. Compression strengths were between 500 N/cm² and 800 N/cm². At that time cores for the production of aluminium, copper, and cast iron parts were made following this procedure.

Simmons reports on progress made in the hot air curing of water soluble water glass powder with module 2.0 in [4.47]. Examined are the conditions of rapid aging and the influence of gassing, air temperature, pressure, flow rate, mould temperature. High strength cores with good surface quality could be achieved by 10 to 15 second gassing with 100 °C to 150 °C air at flow rates of 550 l/min and tools that were heated only to lower temperatures. The warm air-cured water glass had better decay properties than carbon-hardened water glass and the

Figure 4.15: Spray-dried water glass powder

amounts of added binder were very low at 1 % to 1.5 %. Powdered water glass binder systems have the advantage in that low amounts of binder are sufficient to obtain high strength. However, the spread of this variant of the method has been prevented due to the fact that powder binder consist of micro-fine particles. In practical foundry operation it is extremely difficult to keep the work atmosphere free of these particles which can cause irritation of the mucous membranes of the workers. In any case, this was a very interesting approach for improving the properties of water-glass-bound moulding materials.

The water glass powder system is certainly worth exploring as an alternative to the "normal" water glass CO_2 method as it does bring certain benefits. Better flow ability of the moulding mixture and the fact that residual strength can be controlled by low binder content are two advantages that come to mind, for example. The reasons for the cessation of work on the process have been outlined here. It is possible that there may be further development in the future on this method because of the positive aspects involved.

Literature – Capter 4.1

[4.1] Petrzela, L., CSR-Patent Nr. 81931, Anmeldetag 12.12.1947
[4.2] Petrzela, L., Die Erzeugung chemisch gehärteter Formen (CO_2-Verfahren), Freiberger Forschungsheft B 11, 1956, S. 218–267
[4.3] Atterton, D. V., Carbon-Dioxide Process, Foundry Trade Journal 98 (1955), Nr. 2018, S. 479–482, 505–514
[4.4] Böhmer, S., Beitrag zur Theorie des Wasserglas-Formverfahrens, Teil 1, Gießereitechnik 1957, Nr. 1, S. 6–26
[4.5] wie [4.4], Teil 2, Nr. 2, S. 26–32
[4.6] wie [4.4] Teil 3, Nr. 3, S. 49–52
[4.7] Schmidt, W., Philipp, W., Handbuch für das CO_2-Erstarrungsverfahren, Verlag J. Beltz, Weinheim/Berlin 1955
[4.8] Ljass, A. M., Kumanin, I. B. und Krestschanowski, N. S., Hochwertiger Stahlguss, VEB Verlag Technik Berlin, 1955
[4.9] Ljass, A. M., Theorie und Praxis der Verwendung von schnellhärtenden Gemischen mit Wasserglas und Methoden zur Erzeugung von Genaugussstücken mit sauberen Oberflächen, Lit. Proizvodstvo Beiheft 1956, Nr.2
[4.10] Zifferer, L. R., US-Patent 3.318.721
[4.11] Jelinek, P., Vorlesung zu Theorie der Formstoffe, Teil 1, VSB Ostrava, 1970
[4.12] Jelinek, P., Slevarenstvi, 1973; Nr. 12, S. 514
[4.13] Taylor, D. A., Eigenschaften und Prüfung der Formstoffe für das Kohlensäure-Erstarrungsverfahren, Giesserei 1959, Nr. 14, S. 379–399
[4.14] Svensson, I. L., Viscous deformation of carbon dioxide cured sodium silicate, Dept. of Casting of metals, R. I. T, Stockholm, 1984
[4.15] Svensson, I. L., Chemistry of carbon dioxide curing of sodium silicatebinders, Dept. of Casting of metals, R. I. T, Stockholm, 1984

[4.16] Döpp, R., Deike, R., Beitrag zum Wasserglas-CO_2-Verfahren, GIESSEREI 72 (1985), Nr. 22, S. 626–635
[4.17] Gettwert, G., Der gegenwärtige Stand des Kohlensäure-Erstarrungsverfahrens zur Form- und Kernherstellung, GIESSEREI 68 (1981), Nr. 22, S. 659 – 666
[4.18] Gutowski, W., Bledzki, A., Gutowski, P., Technologische Optimierung des CO_2-Prozesses, Gießereitechnik 29 (1983), Nr. 2, S. 39–45
[4.19] Petrzela, L., Slevarenstvi, Nr. 1, 1958, S. 5
[4.20] Ljass, A. M., Lit. Proizvodstvo Nr. 7, 1971, S. 23
[4.21] Srinagesh, K. AFS Int. Cast Metals J. 3, 1979, S. 50–63
[4.22] Renzin, J. R., Sb. Nau. Tr. Politechn. Inta. 94, 1971; S. 163
[4.23] Benda, F. P., Modern Casting 62 (1963), Nr. 4, S. 83 und GIESSEREI-PRAXIS (1974), Nr. 2, S. 35
[4.24] Hespers, W., Kleinheyer, U., Stand der Technik und Kriterien der Entwicklung von Formstoffen in der Bundesrepublik Deutschland, GIESSEREI 65, 1978, Nr. 21, S. 571–578
[4.25] Weiss, R., Kleinheyer, U., Aktuelle Anwendungstrends der Form- und Kernherstellungsverfahren, Altsandregenerierung und Reststoffentsorgung – Teil 1, GIESSEREI 74, 1987, Nr. 21, S. 629–633
[4.26] Firmenbroschüre Gießerei Zlin, CZ
[4.27] Giessereikalender 1990, S. 123–129, Giesserei-Verlag Düsseldorf
[4.28] Flemming, E., Polzin, H., Kooyers, T. J., Beitrag zum Einsatz verbesserter Formtechnologien auf der Basis von Alkali-Silikat-Binderlösungen, GIESSEREI-PRAXIS 1996, Nr. 9/10, S. 177–183
[4.29] Boenisch, D., Knauf, M., Kernschießen – Untersuchungen mit neuartigen Prüfkörpern und verschiedenen Kernbindern, GIESSEREI 78 (1991), Nr. 18, S. 640–646
[4.30] http://www.peak-giesserei.de
[4.31] Polzin, H., Tilch, W., Kooyers, Th., Fortschritte in der Entwicklung des Wasserglasformverfahrens, GIESSEREI-PRAXIS, 2006, Nr.6, S. 169–174
[4.32] Gebauer, A., Gerstmann, O., Das Arbeiten mit wasserglashaltigen Formsanden ohne Kohlensäurehärtung für Grau- und Stahlguss, Gießereitechnik 1957, Nr. 4, S. 73–77
[4.33] Flemming, E., Gisanowitsch, A., Probleme der Zerfallseigenschaften wasserglasgebundener Formstoffmischungen und Möglichkeiten ihrer Beeinflussung durch polymineralische Zusätze, Freiberger Forschungsheft B 179, 1973, S. 127–142
[4.34] Hildebrand, S., Erarbeitung und Erprobung einer Kernherstellungstechnologie auf der Basis modifizierter Wasserglaslösungen für den Leichtmetallguss, Diplomarbeit, Gießerei-Institut, TU Bergakademie Freiberg, 1983
[4.35] Tamarante, A., Untersuchungen zur Aushärtung wasserglasgebundener Formstoffe mittels Warmluftbegasung, Diplomarbeit, Gießerei-Institut, TU Bergakademie Freiberg, 1983
[4.36] Köhler, A., Grundlagenuntersuchungen zur thermischen Härtung von Kernformstoffen auf der Basis modifizierter Wasserglaslösungen, Diplomarbeit, Gießerei-Institut, TU Bergakademie Freiberg, 1983
[4.37] Gettwert, G., Lösliche Silikate, Literaturzusammenstellung anlässlich des 100jährigen Firmenjubiläums der Woellner-Werke Ludwigshafen, Woellner Silikat GmbH Ludwigshafen, 1996
[4.38] anonym, Derivatorische und röntgenographische Untersuchungen von mit Natriumcyclotriphosphat modifizierten Natronwassergläsern, Z. Chem., 1984, Nr. 4, S. 129–130
[4.39] Svensson, I. L., Chemistry and Properties of Sodium Silicate Binders, Part III, On Drying of Sodium Silicate Binder by warm Air, The Royal Institute of Technolgy, Dept. Of Casting of Metals, Stockholm, 1984

[4.40] Carlson, G., Thyberg, B., Core Production by the Sodium Silicate Warm-Air Process, Vortrag Nr. 23, BCIRA International Conference, Coventry/Birmingham 1986
[4.41] Penko, T., Environment Rigth for Silicates Rediscovery, Modern Casting, 1989, Nr. 3, S. 23–26
[4.42] Döpp, R. u. a., Fortschritte des Wasserglas-CO_2-Verfahrens für die Herstellung von Kernen und Formen, VDG-Fachbericht Nr. 051, Düsseldorf, 1988
[4.43] Jelinek, P., Contribution of Czechoslovak Foundry Industry to Chemization of Manufacture of Molds and Cores on the Basis of Alkali Silicates, Slevarenstvi, 1996, Nr. 2, S. 85–103
[4.44] Huusman, O., Lemkow, J., Hot air curing of sodium silicate powder-bonded sand for cores, BCIRA International Converence Coventry/Birmingham, Vortrag Nr. 24
[4.45] Huusmann, O., Das WGP-Verfahren, ein umweltfreundliches Kernherstellungsverfahren, Gießerei-Rundschau 38, 1991, Nr. 11/12, S. 23–31
[4.46] Huusman, O., Lemkow, J., Sodium silicate powder hot-air cured cores, BCIRA International Converence, 1992, Vortrag Nr. 6
[4.47] Simmons, R. E., BCIRA Journal 35, 1987, Nr. 6, S. 409–416 und GIESSEREI 75, 1988, Nr. 21, S. 641

4.2 Cold self-curing processes

In the production of castings in the weight range of about 1000 kg or more, the economically and ecologically interesting methods using clay bonded moulding materials (green sand moulding process) are no more suitable for use due to the excessive thermal and mechanical loads on moulding material for cast metal. In the case of the production of castings in the weight range from 1000 kg up to the largest castings produced in Germany today (between 250 and 300 t), chemically cold self-curing methods or moulding materials are used.

These procedures are characterized by the fact that first a liquid hardener is added to the raw materials. If the binder in the mould of (usually) liquid or solid is added to this mixture, a curing reaction in the form of a polymerization (usually polycondensation, rarely polyaddition) begins. As a result of this hardening reaction, binder bridges form between the individual grains of sand which lead to the strength of the core or mould. This curing reaction occurs normally at room temperature. The cast of the mould thus prepared takes about 24 hours. The final strength of the mould is defined after 24 hours because usually no significant increases in strength are to be expected after that time frame. Since moulds for the production of very large castings by rule cannot be produced within 24 hours, the casting can be held there for a period of several days or even weeks. The hardening of cold self-cured material is well described in the curing characteristics (figure 4.16).

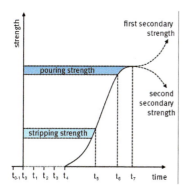

t_{0-1} start mixing time (sand + hardener)
t_0 start of reaction (binder addition)
t_1 end of mixing time
t_2 workability time
t_3 maxim. workability time
t_4 start through hardening
t_5 stripping time
t_6 pouring time
t_7 maximum strength

Figure 4.16: hardening characteristics self-curing moulding materials [3.7], [4.48]

The time frames shown in figure 4.16 describe the key steps in the curing of self-curing material systems. The first time period t_{0-1} means that at this point the curing agent is added to the moulding material. At the beginning of the diagram, at time t_0, polymerization and thus the consolidation of the mould begins immediately after the contact of the binders and curing agent. Upon reaching the mixing time of t_1 the mixed mould material drops into the mould box or the pattern into the core box. At the processing time of t_2, the mould material must be correctly placed in the primary shaping tool and must be sufficiently compacted. It is important to note that further manipulation of the mould material can only be done to the maximum time point of t_3. Although the curing between t_2 and t_3 is already well advanced, mouldings can still be worked within the time of t_3 when they are uncritical or commonly produced castings. If the core or mould is to be used for a difficult casting, it is not advisable for use if it was further manipulated during t_3. The maximum processing time is defined as the point at which a final strength of 75 % of the expected ultimate strength is reached. The reason for this irreversible loss of strength lies in the degradation already present in the early stage of the consolidation of the formed binding structures [3.7].

The next curing characteristic time frame of t_4 describes the onset of complete hardening. From this point the strength increases significantly in the moulding material. Measurable strengths arise due to the rapidly increasing cross linking between the individual sand grains. At time point t_5, the pattern is pulled or the core is removed from the core box. Through the improved open access of the moulding surface to the atmosphere ("open curing"), water evaporation is made more efficient through the binder systems hardened by polycondensation. Thereby, the rate of hardening becomes even faster from this point on. At time point t_6 the cast of the mould is shown and t_7 describes the maximum strength. The further course of the strength development after the casting of the cast part is shown schematically in the curing characteristics upwards and downwards. It has often been the opinion that the residual strength curve for organically-bound moulding material systems basically decreases, while in organic systems the residual or secondary resistance always increases. This interpretation is overly simplistic. Since both inorganic and organic binder systems have a characteristic high-temperature behaviour, the expected residual strength or decay behaviour is depending on these factors:
- Thermal load through the casting material and the resulting pouring temperature
- Removal of the specific mould section from the casting surface and thus the temperature source
- Mass ratio of the casted part and the mass of the casting mould used (metal to sand ratio)

In organic as well as inorganic binder systems (or moulding materials) the problem of decay behaviour/residual strength must always be considered with respect to these three factors.

All cold self-curing moulding methods are dependent on various factors. These factors include temperature (air, mould material, pattern), humidity (air, mould material, pattern) and content of fines.

For the treatment of cold self-hardening moulding material systems it is preferable to use a continuously operating flow mixer that if necessary, can process large amounts of mould material in a relatively short time.

The cold self-curing inorganic moulding techniques currently represent niche markets but will surely be used in broader applications in the future. The following processes are currently available and are described in the following sections:
- Cement moulding process
- Water glass ester process
- Moulding methods with geopolymers

Figure 4.17: Continuous mixer for the treatment of cold self-curing moulding materials (left schematic design, next page mixer in practical use) (images AAGM GmbH/Wöhr)

4.2.1 Cement moulding process

Historical development
The beginnings of the use of cement mould materials for the production of moulds and cores in the foundry industry dates back to the 1930s. One can assume that it is the oldest cold self-hardening or chemically curing moulding process in foundry work. The development of the process began even earlier, at the end of the 19th century. In 1895 in Munich, *Lampel* had already received the patent DEP 88098 for the moulding composition of silica sand, chamotte or cement, and also water glass as a binder [4.49]. What is interesting to note here is the fact that this early version of the process incorporated water glass as the hardener component rather than water. This water glass cement process, which was later widely used, makes an important bypass of a major drawback inherent in the 'classical' water based cement moulding process by significantly reducing curing and hardening times. Long curing time is often cited as one of the main disadvantages of the cement moulding process and is exemplified in low productivity. However, the use of water glass as hardener (as opposed to water) increases the cost of the mould material considerably. A poorer mould material decay is also to be expected. However if one considers all of the improvements in the technological properties of cement mouldings, this older hardening method can again be put in the spotlight.

Further development of the method came about in the 1930s with the use of cement and water as binder. Some important advantages for the production of castings could be achieved by applying lower hydrated mould mixtures. The use of relatively little water leads to a good gas permeability and the incomplete hardening (formation of hardened cement paste) results in acceptable disintegration and stripping properties. These developments were based on patents from *Durand* in the years 1931 (DRP. 520175) and 1932 (DRP. 545123). At this time the method had its largest circulation in France, while in Germany its use was still limited [4.50]. In 1932 there was a paper published in Germany that dealt with the new process and mentioned the interesting possibility of box-less operation [4.51]. After the end of the second World War there was a short-age of raw materials of all kinds, which is why the procedure was again interesting. At that time natural sands were used which caused problems due to their fluctuating composition. Parallel to the turn towards cement sand during this time smaller or serial-cast pieces were shifted from a natural sand moulding process to the (partially) synthetic bentonite-bonded moulding material. Larger moulded pieces and single pieces were shifted from the dry moulding process to the cement moulding process. At the beginning, experiences were drawn from France, England and the USA [4.49, 4.52]. An important contribution to the development of the process in Germany was made by *Gödel*, who conducted successful trials for the production of cast iron parts for the company JE Reinecker in Chemnitz.

The basics of the cement moulding process
The foundation of the cement moulding process is built on the chemical reaction between the reactive components of hydraulic cement and water. As mentioned in Chapter 3, the use of various types of cement is possible. However, the use of Portland cement is preferred because of its favorable strength properties. The production technology and the mechanics of the curing are comparable to the production of concrete in the construction industry. According to *Locher*, the setting and hardening of cement is due to the formation of water-compounds which are formed in the reaction between the cement and the mix water components [4.53]. This reaction is therefore called hydration, while the reaction products that are formed are denoted as hydrates or hydrate phases. Examples of hydration products are calcium hydroxide and magnesium hydroxide, or calcium silicate hydrate.

Where this differs from the construction industry is in the low proportion of water used for sand and cement (moulding material). The cement mould material therefore has a flowable/plastic consistency, which is necessary for mould and core production. When considering the relationship between the cement and water components it should be

a) cement grain before water addition   b) cement grain short time after water addition   c) end of hydration

Figure 4.18: Schematic representation of the hydration of a cement grains [4:58]

noted that the procedure is carried out at low hydration. The cement has less water available than it needs to complete its curing [3.7, 4.54].

Schematically, this process is shown in figure 4.18 of [4.58]. By the addition of water to cement a cement paste is created. Figure 4.18a shows a typical angular cement particle before the addition of water. Shortly after the addition of water the core is surrounded by a shell of fibrous hydration product (cement gel). More water then penetrates the outer gel layer and the hydration of the cement grain continues. The end product is shown in figure 4.18. The complete hydration can take anywhere between several days and several months to complete. This is why the normal 28-day standardization for strength determination of cement has practically no significance for the foundry industry.

Figure 4.19 shows scanning electron microscopic images of solidified binder bridges in a cement mould material. It is clear that the fibrous, needle-like structures are schematically shown and can also be found in the minimally hydrated cement mould material. In the hydration of the cement, shrinkage occurs because the water molecules are incorporated into the lattice of the hydration. This chemically bound water takes up less space than the "free water" so this results in volume loss [4.58]. As already mentioned, the properties of the cement stone formed in the reaction depend largely on the amount of water added. The corresponding performance indicator is the water to cement ratio. Figure 4.20 is helpful in illustrating this relationship. The hydration reaction comes to a halt only when the entire space between the cement particles is completely filled with cement paste. With a water to cement ratio of 0.2 the space meant to be filled with water would likely be filled with cement gel before the hydration is complete. This result would be unhydrated residue in the structure. Figure 4.20 shows the case in which the cement gel is just sufficient to fill the gaps and at a water to cement ratio of 0.4 the water is deposited in the capillary pores [4.8].

Figure 4.19: SEM images of cement binder bridges in the cement moulding material

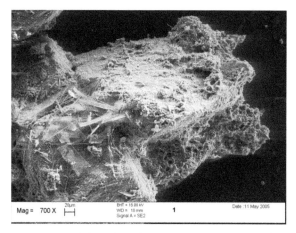

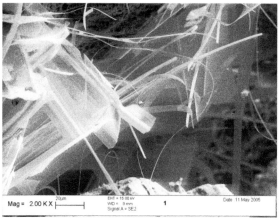

What is apparent from all of this is that it appears that the water in the cement stone is available in different binding states. The resulting cement gel contains chemically bound water, also known as crystal water. The crystal water can be removed only by high temperatures, which would result in the destruction of the hardened cement stone structure. About 25 to 30 % of the cement gel consists of pores. These gel pores are filled with water and always physically independent of the w/c ratio. With a water to cement ratio of 0.4, 40 wt% (of the cement content) is added. After the completion of the hydration within this 40 wt%, approximately 15 wt% are bonded physically in the gel pores and roughly 25 % as crystalline water. If the water to cement ratio is higher than 0.4, the surplus water is imbedded into the capillary pores. They are on average 1000 times larger than the gel pores themselves [4.58].

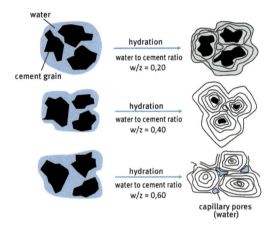

Figure 4.20: Correlation between water to cement ratio and cement hardening [4:58]

In the cement moulding process the transfer to mould and core production means the mould material is used in a highly hydrated state rather than in low hydration state. Therefore it must be assumed that there will be both water in and low gas permeability (in comparison to cement mixtures with minimal additions of water) in the capillary pores. For this reason there must be a separation of the cast piece and the mould after hardening. This is a principle also shared in the use of 'lean concrete' (very low cement content). The reason for operating with a high water to cement ratio is that technologically useful times for moulding, casting and pattern pulling can only be guaranteed by this approach. The correlations discussed here substantiate the fact that due to the water present in the moulding material debris after a relatively short working time, the process is especially susceptible to fluctuations in environmental conditions such as temperature and humidity.

The cement sand moulding method was the preference in the past for the production of larger single pieces, especially thick-walled geometric castings in the mass range of from two to more than 100 t. The preferred application of the method was therefore in general engineering, machine tool castings, and castings for shipbuilding. With the exception of high-alloy steel castings, all cast materials can be cast in the cement sand moulding process. The advantages and disadvantages of the method are presented in [3.7, 4.55 and 4.56].

Benefits of the cement moulding process:
– simple moulding
– reduced compaction effort
– high thermal stability of the mould

- good dimensional stability with large castings
- minimal moulding material costs
- workplace friendly
- very good environmental behaviour
- favorable land fill
- good external usability with accrued sands, for example in road building

Disadvantages of cement moulding process:
- low abrasion resistance/sensitive to erosion
- generally low strength level
- long curing time before being suitable for casting mould
- poor moulding and decay properties
- difficult to unpack and clean
- difficult recycling of used sand

Figure 4.21 shows the basic scheme of the hardening in the cement moulding process. The divisions of response in the short and long-term phases are clearly shown. While in the first phase crystalline hydration products are secreted, crystalline reaction products arise which react very slowly up to the final cured state. Anyone who has ever been involved in the production of concrete is familiar with the long curing times. In addition, the compression strength of cements, which is determined after 28 days, underlines these long periods of time.

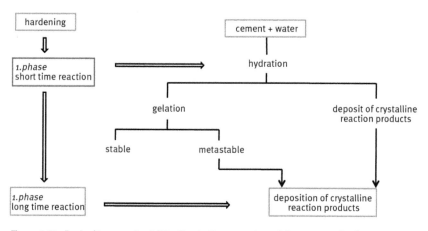

Figure 4.21: Basic diagram of solidification in the cement moulding process [3.7]

Table 4.2: Characteristics of the modified cement moulding process [3.7]

	Cement molasses moulding process	Quick cement moulding process
Development/ application	1965–1975 mainly hand-made moulds and cores	since 1974 mechanized production moulds and cores
Casting mass	Medium size castings (100 – 2,000 kg)	Medium size castings (100 – 2,000 kg)
Composition	8–12 % Portland cement 2,5–4,5 % water 1,5–3 % molasses 85–80 % silica sand 0–0,5 % curing accelerator (Fwd. $CaCl_2$)	5–8 % quick-setting cement 3–6 % water 0–0,1 % surfactants 85–80 % silica sand
Pouring time	after 24 h	after 6 h

Therefore the main problems with the cement moulding method is the long stripping and hardening time of 36–48 hours and the bad mould material decay after casting. Therefore, work in the technology of cement moulding technique was undertaken early in the development. In the course of this work, the cement molasses and rapid cement moulding process emerged. The main characteristics of the modified mould material method are shown in table 4.2:

The preparation of the moulding material is analogous to the procedure for other cold self-hardening mouldings. Silica sand is usually used as a moulding material with an average grain size 0.30–0.60 mm. However, the use of any other form of raw materials is also possible. The use of relatively coarse-grained forms of raw materials is essentially based in the fact that the method is now used only in very few foundries to produce castings in the mass range from 50 to over 100 tons. The moulding sand preparation is carried out in continuous-flow mixers. Then, the moulding material is premixed with the cement when it's still dry [4.54]. The cement content in the moulding material is 7–11 % according to figure 3.7. The weight ratio between water and cement added is referred to as the water cement ratio. A ratio between 0.6 and 0.7 is considered beneficial. Accordingly, the amount of water added depends on the cement content. Depending on the water cement ratio, the proportion of water in the mould is between 5 and 7 %. The addition of following form material additives or accelerators is also possible:

- Coal dust (2 – 4%) → improved setting process but worsened sintering reaction
- Organic binder additives (Dextrin, starch) → increased green strength and favorable decay properties
- Accelerator (NaCl, molasses) → acceleration of setting time

The production of the moulds and cores is generally carried out by manual fabrication. The moulds are usually produced as a stove or pit shape. The production of large box shapes would also be possible, but is prevented because of the poor decomposition properties of the moulding material. This is usually worked with backfill material. Therefore, the facing mould material needs to be of high quality (high fire resistance and gas permeability). The backfill material has "only" a support function. However, it should still have good gas permeability and good disintegration characteristics after casting. Lower quality silica sand or even reclamation can be used. In addition, reduced binder content can be worked with which further reduces material mould costs [3.7].

Favored by the current debate over the increased use of inorganic binders systems and mould processes based on them, the cement process should once again become the focus of interest. The technological characteristics of the classic cement moulding process described so far have led to a justified reduced use of its components, because significantly better functional and technological properties are seen first in other inorganic systems (e. g. the water glass-ester process) and later in processes based on organic binder systems. Particularly, the outstanding advantages of the process that make it economically attractive, and the ecological safety inherent in the method might be cause for it to be considered for mainline use again. It might be possible also to use the cement moulding process in conjunction with the compression moulding process to make it more feasible from a business standpoint.

The following approaches for improving the process are worth considering:
1. The use of higher quality cement, such as alumina or special cements
2. Special hardening components for the binder material such as the use of the water glass cement method which is not presently in general use
3. Use of accelerators to increase productivity
4. Investigation of different qualities of the Portland cement binder system which is the cheapest alternative system

Two of these approaches will now be discussed in more detail.

The use of high quality Portland cements

To improve the strength properties of cement mouldings, the alternative to CEM I 42.5 R has currently been CEM I 52.5. This cement is characterized by the higher strength designation of N/mm² after 28 days. This is achieved mainly by a higher fineness of grind. The quality of the material is reflected in a higher price.

To determine the strength properties of Portland cement CEM I 52.5 R a series of experiments were performed with modified cement content. For example, the shear and bending strengths of the corresponding mould mixes or moulded plastic test specimen were determined. Starting with 8.5 wt % addition of cement to the moulding material it was then dropped by one-percent increments down to to 6.5 wt%. Since reducing the cement content changes the water to cement ratio, test series were also examined with a modified water and cement content. Water to cement ratios arose with values of 0.7, 0.8 and 0.9 along with the following polymer formulations (table 4.3) [4.7]:
- CEM I 52.5 R 8.5% cement content with w/c ratio of 0.7
- CEM I 52.5 R 7.5% cement content with w/c ratio of 0.7 and 0.8
- CEM I 52.5 R 6.5% cement content with w/c ratio of 0.7, 0.8 and 0.9

The results were compared with the corresponding data for the reference mix with the cement CEM I 42.5 R at a binder content of 8.5 % and a w/c ratio of 0.7.

In [4.57] it is shown that with the higher quality cement CEM I 52.5 R containing 8.5 wt% a higher strength advantage is gained over CEM I 42.5 R cement. An initial strength advantage (after 48 h) of 50 %, according to DIN 197-1, could only partially be detected. However the CEM I 52.5 R samples reached strength level after 36 hours while the CEM I 42.5 R samples reached it after 48 hours. This means that the use of this recipe of cement moulding material would be possible by applying acceleration methods. Figure 4.22 confirms this statement when the binder content is 8.5 %. Clearly visible is the increase in strength when switching from cement 42.5 R to 52.5 R, while both exhibit the influence of ambient temperature on the acceleration of water extraction. As a result of the higher fineness of the cement 52.5 R, the gas permeability of the mould material is deteriorated. This is shown in figure 4.23. It should be noted here that the measurement of gas permeability was measured indirectly through the determination of the dynamic pressure (high back pressure – low gas permeability).

It should be observed here that the final strength of the mould material recipe is greatly increased. Furthermore, it will also result in higher mould material costs. There are advantages in strength development due to the increased hydration of CEM I 52.5 R compared to CEM I 42.5 R. With a gradual reduction of the cement content in the CEM I

Table 4.3: Investigated mould material formulations [4.61]

Moulding mixtures (cement)	CEM I 52,5 R 8,5 %			CEM I 52,5 R 7,5 %			CEM I 52,5 R 6,5 %		
w/c-ratio	0,7	0,8	0,9	0,7	0,8	0,9	0,7	0,8	0,9
Cement content [wt.-%]	8,5	–	–	7,5	7,5	–	6,5	6,5	6,5
Water content [wt.-%] (target moisture)	5,3	–	–	4,7	5,3	–	4,1	4,7	5,3

52.5 R samples to 7.5 wt%, the strength values were adjusted to match the CEM I 42.5 R samples. That the strength levels achieved are dependent on moisture content was confirmed at least in the upper strengths of the CEM I 42.5 R samples. It proved to be advantageous to adjust the moisture content of the ambient temperatures. That is, low temperatures yielded a low moisture content (w/c ratio = 0.7) and high temperatures yielded a relatively high moisture content (w/c ratio = 0.8) for optimum strength. For production it is recommended that moisture content should be adjusted according to the season of the year. The calculated and compared gas permeability in the sets of samples had changes in the corresponding areas of the reference samples (figures 4.44 and 4.25) [4.61].

The identified strengths of CEM I 52.5 R series of samples (cement content = 6.5 wt% corresponded to the lower level of the corresponding CEM I 42.5 R reference sample series. Therefore, the substitution of CEM I 42.5 by the CEM I 52.5 can theoretically yield an optimal cement content of 7.0 wt % for the mould material to be used. The corresponding water content should be adjusted according to the data gleaned at the ambient temperatures [4.57]. These results show that where previously only CEM I 42.5 cement could be used, it could be in principle, replaced with the better quality CEM I 52.5 R [4.59]. A faster strength development with higher ultimate strengths can then be applied to speed up the moulding production and thus increasing the productivity. To offset the price differences between the two cements, reduced amounts of binder can be used when working with CEM I 52.5 R cement. The positive side effect would be lower residual strength and better disintegration behaviour. As an additional processing parameter, water content can also be varied as a function of the ambient conditions. This can be done within the required processing time frame. It is important to take into consideration when using CEM I 52.5 R, particularly at higher water to cement ratios (or even the development of other hydration products) a deterioration of gas de-

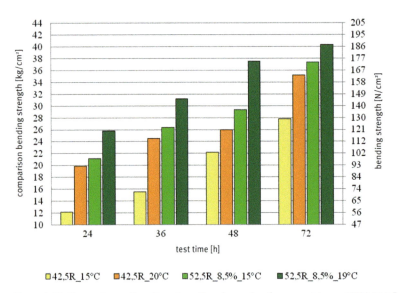

Figure 4.22: Comparison of bending strengths when using the cement types CEM I 42.5 R and CEM I 52.5 R with a cement content of 8.5% [4.57]

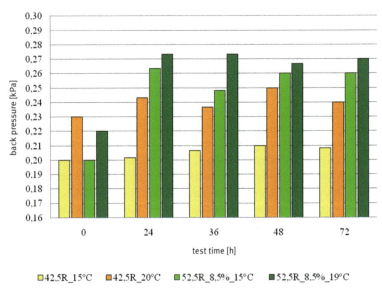

Figure 4.23: Comparison of gas permeability in the cement types CEM I 42.5 R and CEM I 52.5 R with a cement content of 8.5% [4.57]

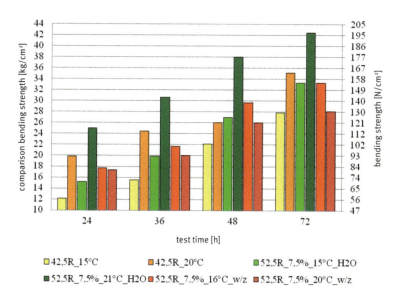

Figure 4.24: bending strength strengths of cement CEM I 52.5 R at 7.5 % binder content compared to the cement CEM I 42.5 R at 8.5 % binder (H_2O means additional w/c -0.8) [4.57]

velopment happens. To counter this there are established steps such as changing the grain belt used for the mould material.

The use of additives in the cement moulding process
It has been repeatedly stated that the traditional cement moulding process has some pretty serious drawbacks. One of the most important of these is the slow curing of the moulding mixture to speed up the curing one can incorporate general suggestions commonly known in concrete technology. In addition to concrete plasticizers (CP) and solidification retarders (SR), hardening accelerators (HA) are used. Hardening accelerators have a chemical effect on the reaction mechanism between the cement particles and the mixing water. The result is a faster blending of the cement components and intensification of the hydration. This increase in the rate of hydration is based on a strong development of hydration heat. Accelerators used in the construction industry today contain organic substances as well as sodium or calcium compounds. In figure 4.26 the effect of a hardening accelerator and a retarder is shown in comparison to "normal" hydration with water.

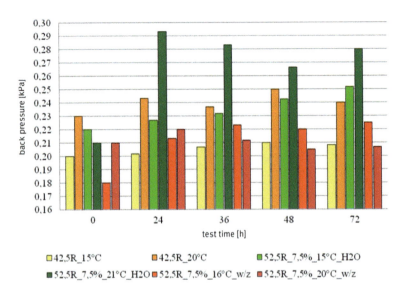

Figure 4.25: Gas permeability of cement CEM I 52.5 R at 7.5% binder content compared to Cement CEM I 42.5 R at 8.5% binder (H_2O means additional w/c -0.8) [4.57]

The use of hardening accelerators, the basis of $CaCl_2$ for example, does not affect the hydration phases formed during cement hydration but shifts the pore distribution (in the concrete!) to smaller pore diameters (gas permeability). The early strength is increased but then is diminished after prolonged setting time. This can be explained by a shift in the axis rations of the Portland crystals. Compliance with exact dosage is important when using accelerator [4.60, 4.61 and 4.62].

For example, recent studies deal with the use of calcium chloride as an accelerator. In [4.63] a commercially available additive is investigated on the basis of possible achievable property improvements. Table 4.4 shows the mould material mixture compositions used in the experiments.

Figure 4.27 shows the effect of the accelerator on the compressive strength of the moulding mixture. It is clear that this additive may actually be an accelerator for the cement moulding method in the classical sense. Even at the first strength measurement taken after 20 hours, the difference is clearly shown from the mould material mix without accelerator. This difference continues until the end of the measurement after pat-

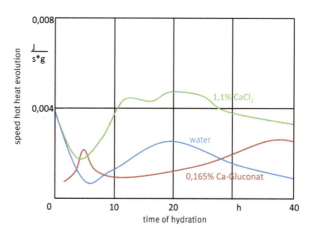

Figure 4.26: Influence of an accelerator ($CaCl_2$) and a retarder (calcium gluconate) to the heat development during hydration of Portland cement [4.60]

tern hours which is the normal time to the pattern extraction and the strength is about 150 % compared to working without additive.

Figure 4.28 represents the measured values in gas permeability of the mould material mix. It is noted that the gas permeability is slightly lower with the use of the accelerator as a result of quickly forming hydration products. The crystals grow faster in the pore spaces of the mould material mix when using accelerants. Although not implicitly stated in the document [4.63], it is likely that there is less thermal expansion in the mould material. Therefore, the stability of the mould wall must be considered. However, in this regard there should be an improvement of behaviour when using accelerators.

In the course of these studies it is shown that the use of hardening accelerators containing calcium chloride can cause a significant increase in gas development. If such a product is still to be used, adequate measures for gas exhaustion are necessary for the mould material.

Table 4.4: Form of composition experiments with accelerators based on $CaCl_2$ [4.63]

Mould component	Dosage
Silica sand AGS 0.46 mm, GG 69 %	100 parts
Cement CEM I 42.5 R	1 part
Water	0.5 parts
Additive basis $CaCl_2$	3 %, based on the cement content

Cold self-curing processes | 85

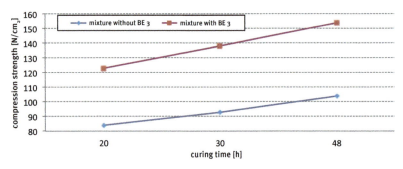

Figure 4.27: Compression strength of cement sand mould materials (table 4.4) with and without the use of CaCl$_2$-based additives [4.63]

With other inorganic accelerators deals [4.64], specifically with special silica sols. Table 4.5 shows moulding mixtures with two different accelerators added at between 0.5 and 1.0 % of total cement content. The experimental program in table 4.6 is based on certain test times and outlines where the dimensions of these times are in the general cement moulding process. In addition to the strength, gas development and permeability are considered. Finally, the decay behaviour in relation to compression strength is also examined.

As a result of the work, the effectiveness of the two accelerators on the development of strength could be demonstrated unequivocally. For the test series with the CEM I 42.5 cement a huge increase in pressure and bending strength is observable which relates to the overall strength in both the early and late stages. In the experiments with cement CEM I 52.5, an increase of the 10 hour and 24 hour compressive strengths are

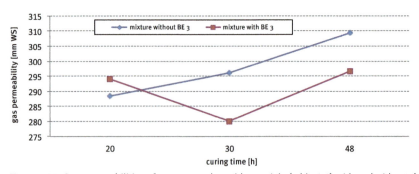

Figure 4.28: Gas permeabilities of cement sand mould materials (table 4.4) with and without the use of CaCl$_2$-based additives [4.63]

Table 4.5: Mould material composition studies on the use of curing accelerators for cement moulding process [4.64]

Composition of the mould material mixture	Weighing
Mass parts 100 silica sand H 32	11.000 g
10 mass parts of Portland cement CEM I 42.5 R or 52.5 R	1.100 g
6 mass parts of water	660 g
0.5 mass% accelerator based on cement	0.5 % = 5.5 g
1.0 mass% accelerator based on cement	1.0 % = 11 g

found. The late compression strengths have lower values compared to pure cement mould material. However, there were no significant changes in the bending strength except for a minimal increase detectable only at 10 hours.

For gas permeability, an opposite picture is shown when using accelerants. In the experimental batches of CEM I 42.5 the gas permeabilities fell off significantly. For mould mixes with cement CEM I 52.5 they went up slightly. The reason for this behaviour is evident in the altered reaction chemistry (water separation). In regards to residual compressive strength, the results are also in opposite directions. The addition of the accelerator in the mould materials can influence decay behaviour. The residual compression strength increases with the use of CEM I 42.5 cement while it decreases moderately with use of CEM I 52.5 cement. An influence of the accelerator on the amount of gas evolution could not be established conclusively. Generally speaking however, increased gas evolution is to be expected with the use of accelerators based on the larg-

Table 4.6: Test program for evaluating the accelerators [4.64]

Type of test	Time of tests				
	10 h	24 h	36 h	48 h	72 h
Compressive strength	X	X	X	X	X
Bending strength	X	X	X	X	X
Gas permeability		X		X	X
Residual compressive strength					X
Gas evolution					X

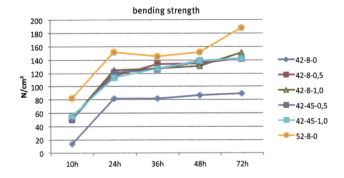

Figure 4.29: Bending strength when using cement CEM I 42.5 last two numbers denotes accelarator component type and content [4.64]

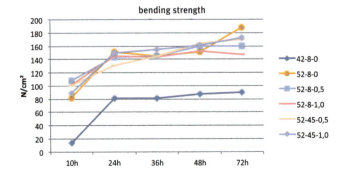

Figure 4.30: Bending strength when using cement CEM I 52.5 last two numbers denotes accelarator component type and content [4.64]

er amount of water in the system. Gas quantities are not critical here in terms of technological impact, except in the risk of gas bubbles.

Summary of the cement moulding process

As mentioned, the cement moulding process is probably the oldest method of chemically curing processes within the foundry industry. New inorganic binders systems as well as moulding methods based on organic binder systems (which since the 1970s had led to improvements in the application stage) almost entirely replaced the cement moulding method in the foundry practice.

The manufacture of ship propellers described here earlier, are Germany's absolute exception. But if one were to observe which cold self-curing systems are in use today, only the procedures described in the following two chapters would be found. That is, the water glass ester process and methods using so called 'Geo-polymer' binders. The cement moulding process has several drawbacks such as relatively slow curing and low strength, but it also has some important advantages. To exploit these advantages, such as cheap materials or safety of the used sand, further method developments are needed. The examples presented here demonstrate that such improvements in the technological properties are quite possible. Although the cement moulding process is not going to be generally applied in the foreseeable future, it certainly could re-conquer some applications in the cold self-curing processes. Because this document is concerned only with moulding materials and moulding methods, ultimately one should refer to a publication by *Tomek* [4.65] among others. Recent literature on the cement moulding method is rare. The authors of the above named article/s are concerned with the influence of cement based cores on the microstructure and mechanical properties of cast iron materials.

4.2.2 Water glass ester process

The water glass ester process is used for making moulds and cores of wet, pourable, moulding materials. This document is written on the subject of inorganic binder systems, therefore, it should be noted here that although the method uses an inorganic binder, an organic hardener is used. However, in contrast to the completely organic based cold self curing systems, the water glass ester process uses a much smaller amount of hardener at 8–12 % based on binder content. Furthermore, the hardeners are harmless compounds, such as esters of acetic acid, which release only a few harmless emissions during the pyrolysis of pouring.

Even before the development of the water glass ester process, which draws on the work of the Foseco Company, there was a series of cold self curing form methods on the basis of the water glass binder, especially in Eastern Europe and the former Soviet Union. In [4.66] the situation in the former GDR is described by *Schuster*. In addition to the cement moulding process already described in detail, the following variants of the water glass moulding process were known and in use:
- the water glass ferrochromium slag process (ferrochrome slag is a waste product of metallurgy)
- the water glass ferrosilicon process (ground ferrosilicon, the so called Nishijama method)
- the water glass cement process
- the water glass clay process

Due to the current practical insignificance of these methods, a detailed presentation is omitted here. However, it remains unclear whether these variants could provide start-

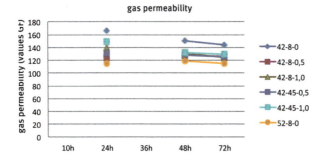

Figure 4.31: Gas permeabilities using cement CEM I 42.5 last two numbers denotes accelarator component type and content [4.64]

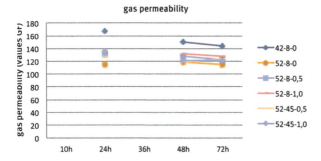

Figure 4:32: Gas permeabilities using cement CEM I 52.5 last two numbers denotes accelarator component type and content [4.64]

ing points for future developments in the field. In [4.67] it is was determined by *Borsuk* and *Vasileva* that for 1976, "Self-curing water glass mixtures with liquid hardeners from the class of complex esters such as diacetin, triacetin and ethylene glycol diacetate" will be produced. Furthermore, the authors noted that water glass ester moulding mixtures have a high fluidity and easily yield to compaction. Among the other good processing properties it is noted that the pattern can be extracted after only 40 minutes. An interesting contribution is from *Beckius* in the same journal [4.68] that deals with the use of olivine sand as a moulding material. The use of alternative moulding moulding materials will be discussed elsewhere in this publication. *Beckius* here is reporting on the attempts to produce gray iron and cast steel. The mould materials used contained 3–5 % binder and therefore 10 % of curing agent, based on the binder content. Moulds and cores at comparable strengths were achieved as furan resin mould material with silica sand.

Schumann and *Lindemann* introduced a water glass ester mould material system under the name of Gisacodur. The binder of this system included the normal practice of using the water glass CO_2 process and organic disintegration promoters. The authors note the scope of this primarily in the mould and core production for medium size castings and products made of cast iron and lamellar graphite. *Schumann* and *Mai* go into detail on this system which combines the advantages of inorganic and organic binders [4.70]. Through a wide range of applications, the method was therefore able to replace water glass clay, glass water CO_2 and phenolic resin moulding materials. The solidification and curing characteristics can be controlled within wide limits by the use of different curing agents. Processing times are represented by timeframes of between 5 and 60 minutes while model extraction times range between 30 and 150 minutes. *Gerstmann* [4.71] also recommends a binder content of 3 % to 0.3 % hardener component which means a favorable ratio between primary and secondary strength is ensured.

Burian and *Kristek* pointed out that when incorporating hardening by water glass binder through carbon dioxide approximately only 10 % of the potential strength can be realized [4.72]. This in turn makes necessary the use of higher binder content, thereby forcing a worsening of the decay behaviour. The authors still refer to the need for the increase in strength and other property improvements in regards to self-hardening mould mixtures. To evaluate the properties of the water glass it is recommended to examine the silicate modulus, density, viscosity, and to determine the coagulation

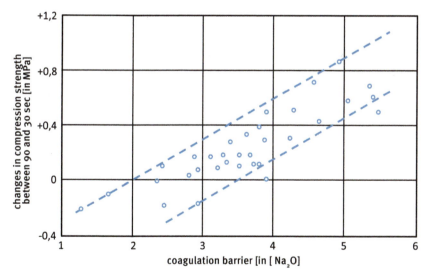

Figure 4.33: Cure profile of water glass ester moulding materials with 3 wt% binder and 0.3 wt% hardener for use of binders with different coagulation barrier [4.72]

barrier. Through the coagulation barrier the electro-kinetic ratios in the colloidal system of the water glass are determined. If an electrolyte is added to this colloidal solution, the number of ions in the dispersion medium of the solution increases. The electrokinetic potential is reduced by an adsorption exchange between the outer double layer and the electrolyte. At some point, the adhesion forces gain the upper hand over the electrostatic repulsion and coagulation occurs in the solution. The minimum amount of electrolyte necessary for coagulation is referred to as the coagulation threshold. HCl for example, can be used as an electrolyte, and through the conversion of the HCl content on the (to be neutralized) Na_2O, one can directly specify the content of the stabilizing ions in the double outer layer of the micelle. With an increase in coagulation threshold there is also an increase in pressure strength and decay behaviour therefore, one can expect an improved storage capacity for the produced parts. As

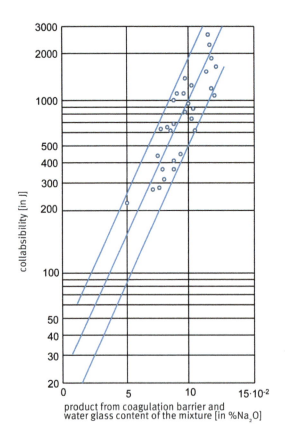

Figure 4.34: The collabsibility of water glass ester sands depend from the coagulation barrier if the used binder solution [4.72]

Table 4.7: Calculation of the equivalent quantities of hardener according to the coagulation barrier and derived processing time [4.72]

	Binder hardener system				Sand with water glass binder and equivalent amount of hardener				
Sample-No.	Coagulation barrier % Na$_2$O	Equival. hardener amount in % binder	Sped gel formation in min		Binder amount equal		Hardener amount equal		
			initial	final	Binder in kg/f 100 kg sand	Processing time min	Binder in kg/f 100 kg sand	Processing time min	
1	2,89	7,5	10,5	16,0	3	10	4,16	10
2	3,41	8,8	9,3	17,0	3	13	3,53	10
3	4,00	10,4	9,7	17,5	3	13	3,00	13
4	4,46	11,6	9,8	17,5	3	13	2,69	14
5	4,92	12,9	9,8	17,5	3	13	2,44	13
6	6,83	17,8	9,3	15,5	3	12	1,75	11

Cold self-curing processes | 93

shown in pictures 4.33 and 4.34 these parameters can also be used to an advantage in the use of self-hardening water glass ester mould materials. Through the coagulation barrier the reactive characteristics can be profiled and the hardening curve can be predicted. Furthermore, it is possible to calculate the amount of hardener required for the gel formation. There is evidence that the processing time of the moulding material ("Bench life") at different binder contents remains practically the same when the equivalent binder to hardener ratio is set (table 4.7).

In [4.73] *Jelinek* deals with the self curing version of the popular water glass CO_2 process which at the time represented the most important cold self hardening moulding of technology in the former Czechoslovakian Republic (and also in the states of the former Soviet bloc). *Jelinek* pointed out in 1979 that inorganic binder water glass is between 10 and 20 times more cost efficient and much more environmentally friendly than organic binder systems. *Jelinek* also reports that due to the attainable strengths, binder contents of between 2 and 3 % are feasible. The solidification reaction in the process proceeds with a shift of the pH value by the formation of an acid to the formation of a salt (sodium acetate CH_3COONa) as shown in figure 4.35. Actually, the desired product of the reaction are $Si(OH)_4$-gels which achieve strength in the composite particle. According to *Flemming* and *Tilch* [3.7] the solidification of the mould material mix occurs through the hydrolysis of the ester in several stages up to an equilibrium state where the water is stored within the colloidal silica solution. The liberated acetic acid lowers the pH and thus the electrokinetic potential as the $Si(OH)_4$-gels separate. During solidification a strengthening effect in the material composite is also attributed to the outgoing sodium acetate. Figure 4.34 shows the processes involved in this consolidation method. *Jelinek* further stated in [4.73] that from an occupational hygiene, technical and economic point of view, the esters of acetic acid monoacetin,

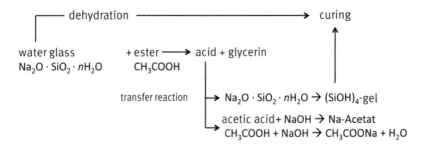

Figure 4.35: The process of the curring in the water glass ester process [3.7]

diacetin and triacetin as well as the esters of ethylene glycol, ethylene glycerol diacetate and diethylene glycerol diacetate prevailed as a hardener. Currently, other esters are used as hardener which is a subject to be discussed later in this document.

The esters differ in their rate of hydrolysis. While monoacetin reacts very fast, the reaction with triacetin takes place in three steps and is thereby, substantially slower. This makes it possible to react to fluctuating environmental conditions during moulding sand preparation, such as temperature and humidity, by using mixtures of various esters in varying ratios. As *Jelinek* described, the stepwise hydrolysis of the triacetin through the formation of diacetin and monacetin ends in the formation of acetic acid and glycerol. The acetic acid reacts with sodium hydroxide, and sodium acetate is formed.

$$CH_3COOH + NaOH \rightarrow CH_3COONa + H_2O$$

The compound of sodium acetate crystallizes afterwards, and forms needle-shaped salt crystals which pass through the gel. While *Jelinek* believes that these needles solidify the gel material, other authors suggest a weakening of the binder bridges through the internal notch effect of these needles. While the controllability of the curing process by use of different hardeners is undoubtedly an advantage, the temporary formation of glycerol must be noted as a disadvantage. This drawback in water glass-ester mould materials is known as temporary plastic behaviour. Jelinek also presents a method to determine the reactivity of the ester hardener [4.73]. For this he regards in parallel the flow times of a defined binder hardener mixture out of a 6 mm flow cup and the penetration time of a needle into the solidifying gel as the temperature increases during curing.

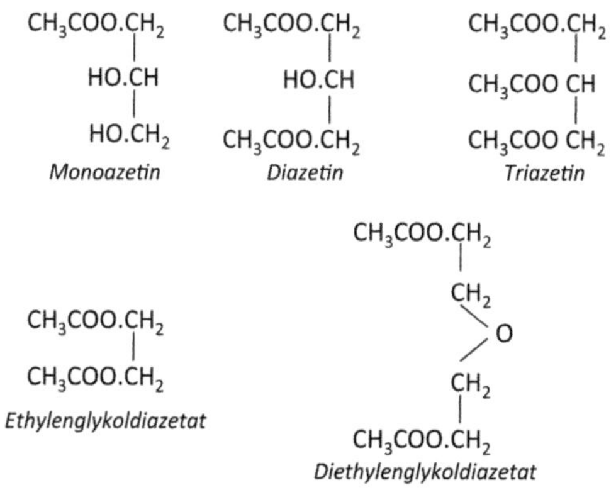

Figure 4.36: Often used ester as hardener in the water glass ester process [4.73]

$CH_3COO.CH_2$
$|$
$HO.CH$
$|$
$HO.CH_2$
Monoazetin

$CH_3COO.CH_2$
$|$
$HO.CH$
$|$
$CH_3COO.CH_2$
Diazetin

$CH_3COO.CH_2$
$|$
$CH_3COO\ CH$
$|$
$CH_3COO\ CH_2$
Triazetin

$CH_3COO.CH_2$
$|$
$CH_3COO.CH_2$
Ethylenglykoldiazetat

$CH_3COO.CH_2$
$|$
CH_2
$\diagdown$
O
$\diagup$
CH_2
$|$
$CH_3COO.CH_2$
Diethylenglykoldiazetat

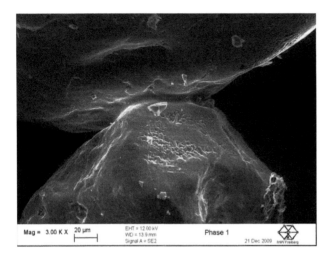

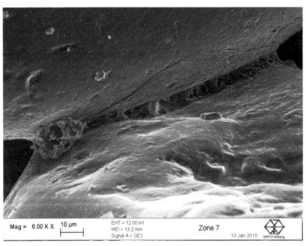

Figure 4.37: Homogeneous (top) an inhomeneous (bottom – through sodium acetate) binder bridge in the water glass ester process

Apart from the ester, solidification in water glass ester process, as well as in all water glass processes, is affected by the molar ratio of the silicate binder solution. While a higher modulus causes a faster cure with lower final strength, lower modulus causes a slower strength increase but higher ultimate strengths (figure 4.38). The method, like all cold self-hardening moulding methods, has a sensitivity to temperature variations. As a general rule it can be considered that at a temperature fluctuation of +/-1K the processing time changes to 1.5–2 minutes. The most favorable temperature range for the processing of the moulding material is regarded as 10 °C to 25 °C. water or alcohol based coatings can be used for the surface treatment of the mold piece. Pattern materials can be wood, plastic or metal. For mould production, 3–3.5 % of binder is suggested, while in core production 3.5–4.0 % binder content is necessary. For the content of the hardener, approximately 12 to 14 % are specified relative to the binder [3.7].

Jelinek shows inter alia in [4.7] that in the area of self hardening variants for water glass processes, further improvements can be achieved by the exotic method of magnetic treatment of the binder (for example). It is thus shown that a 25 % extension of the processing time can be achieved through magnetic field treatment. Even with prolonged processing time, the curing of the moulded parts is achieved faster. For example, strength is 20 % higher with 30 seconds of magnetic therapy after 1 hour of curing time. Despite the improved strength properties there is no worsening of decay properties or detectable negative effects on the casting surfaces. Figure 4.39 shows an example of the achievable strength increases in the use of magnetic treatment.

Svensson looks at decay behaviour affecting all water glass processes after casting in [4.75]. In his investigations he uses a binder content of 4.0 % (water glass modulus 2.7) and a hardener content (mixture of 60 % diacetin, 40 % triacetin) of 0.4 %, and determines the shear strength under various environmental conditions. He notes that the solidification by ester is more susceptible to changes in property changing environmental conditions compared to the drying of water glass bonded mould materials. Further-more, the influence of the following additives is considered on the residual strength: 10 % sugar compound (sorbitol), and 25 % MgO and 25 % Mg(OH)$_2$. The effect of these additives is made clear in figure 4.40. Where as the addition of sugar results in a negative effect on both the first and second maximum strengths, the addition of the two inorganic additives brings with it the reduction of either one or both maximums. Therefore, the samples are exposed to one hour of the inspection temperature and tested

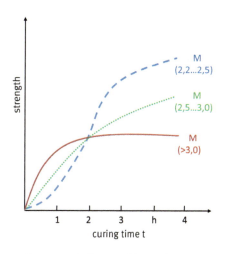

Figure 4.38: Influence of the modulus at the curing of water glass ester sands [3.7]

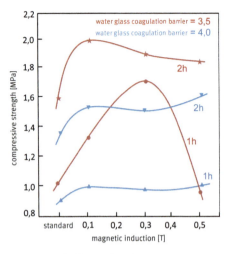

100 MT silica sand 0-36
3 MT water glass binder
0,3 MT hardener

Figure 4.39: Compression strengths of water glass ester sands before (blu) and after (red) a magnetic field treatment [4.7]

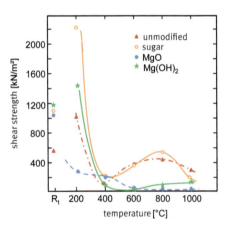

Figure 4.40: Residual shear strengths of water glas ester sands with different additives [4.75]

after cooling to room temperature. To explain the observed phenomena *Svensson* uses scanning electron microscopy to assess the appearance of the fracture surfaces on the binder bridges in the core group.

In examining the binder bridge break surfaces of the unmodified water glass binders at a temperature load of 200 °C he finds that, due to the increased drying of the binder film, the binder bridge is as homogenous as it is at room temperature. When the temperature increases to 400 °C, round holes in the binder bridges (which are due to the evaporating crystal water) form. This phenomenon is amplified at a temperature of 600 °C whereby the binder bridges have more holes. According to *Svensson*, this mechanism may explain the fact that water glass bonded moulding materials exhibit the familiar strength minimum in the range between 300 °C and 600 °C. Therefore, at a temperature of 800 °C the glass phase formation starts and the associated increase in strength. At 1000 °C the first cracks are formed due to stresses in the bridge and thereby a decrease in strength begins. If the three mentioned additives are added to the binder, the strength behaviour shown in figure 4.40 is realized. *Svensson* explains the fundamental course of strength at elevated temperatures with the described mechanisms of water evaporation and the cracking due to stress development. With the addition of the inorganic additives the morphology of the binder bridges is affected. Magnesium initially takes on water and forms magnesium hydroxide, and thereby the first drop in maximum strength results at 100-200 °C from the disabled or slowed drying. When using the additive $Mg(OH)_2$ no water is adsorbed and as a result the strength increases at 200 °C by drying. At temperatures between 400 °C and 600 °C the hydroxide in both cases loses crystal water which results in lower strength. Noteworthy in figure 4.40 is that with the addition of sugar, no significant strength reductions are apparent.

Flemming and *Tilch* summarized in [3.7] that the water glass ester process is suitable for both mould and core production. They characterize the method by the following advantages:
- by using different hardeners, processing characteristics can be adjusted within wide limits
- the good flowability of the mould material allows the production of difficult cores
- the mold material is eco-friendly

Conversely, a number of disadvantages of the method are also determined:
- the strengths (in this case bending strength) are lower than with cold resin methods
- in cases of insufficient thermal stress, bad mould material decay can be expected therefore a casting wall thickness of at least 10 mm is recommended
- the mould material is thermoplastic in certain temperature ranges which can lead to cavities in the form and poor recycling of used sands.
- the abrasion resistance is worse than with resin bonded moulding materials

Flemming and others introduce a group of advanced binder systems with improved properties during a very difficult acceptance phase for water glass methods [4.76]. While the core binder systems used in this article were modified with different additions of organic additives, the mould binders which were specifically designed for self hardening process variants, contain only inorganic and to a lesser extent, organic modifiers. These systems achieve, for example, bending strengths of 180 N/cm^2 after only 8 hours of curing. This is likely to be significantly exceeded after 24 hours. Also interesting here is the representation of residual strengths (figure 4.41). While the inorganic modification is only achieved in the area of second maximum strength, the strength-reducing effect of organic additives is clearly shown here.

Sachurak inter alia deals intensively with the ester hardeners used [4.77]. In order to evaluate the reactivity or expected reaction speed in hardening they introduce the ester number. The principle is based on the alkaline hydrolysis of the ester when heated in water or alcohol. As a result of the reaction, alcohol, as well as a corresponding salt of carbonic acid, is formed. Through this method the "ester number" or "saponification" is determined. This ester number is therefore equivalent to the amount of sodium hydroxide (in mg) which is necessary for the saponification of all the esters present in 1g of the substance to be analyzed. The method is based on the titration principle and can be applied quite easily. Table 4.8 shows 5 esters that are applicable for this method along with their ester numbers and a description of curing behaviour. Esters with the largest number of unsubstituted acid radicals are more reactive. If one incorporates the common practice of mixing two esters with different degrees of substitution of the OH groups, the ester number and thus the reactivity of this mixture can be determined.

The ester number is significant for about 8 to 9 hours. After 24 hours of hardening this picture is reversed into its opposite: The hardeners with high ester value have higher

Table 4.8: Typical hardeners for the water glass ester process, correlation between ester number and the curing speed [4.77]

ester	chemical formulation	average ester number mg NaOH/g	qualitative characterisation *
glycerinmonoacetat (monoacetin)	CH_2COOCH_3 | $CHOH$ | CH_2OH	ca. 300	very fast (below 1 min)
glycerindiacetat (diacetin)	CH_2COOCH_3 | $CHOH$ | CH_2COOCH_3	ca. 450	fast (3 to 4 min)
glycerintriacetat (triacetin)	CH_2COOCH_3 | CH_2COOCH_3 | CH_2COOCH_3	ca. 550	very slow (more than 120 min)
ethylenglycol-monoacetat	CH_2COOCH_3 | CH_2OH	ca. 400	very fast (below 1 min)
ethylenglycol-diacetat	CH_2COOCH_3 | CH_2COOCH_3	ca. 550	slow (90 min)

* qualitative characterisation of the ester through the interaction with the water glass

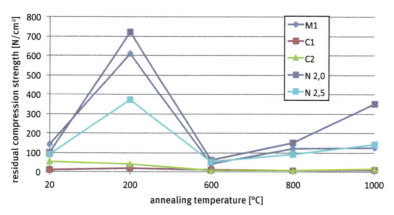

Figure 4.41: Residual compression strengths of different modified water glass binders with ester curing (3.0 % binder; 0.3 % hardener) [4.76]

strengths. This principle highlights the well known fact that when using a faster hardener, lower ultimate strengths can be expected. The practical significance of this varies on a case by case basis. For the application of the achieved insights the nomogram (figure 4.42) included in [4.77] can be used, which shows the connections between ester number, compression strength and curing time.

Jelinek notes in [4.78] that esters with liquid glycerols form during the solidification of water glass binders. He examines inter alia the dependency of the expected strength on the particle size of the sol particles present in the binder solution and refers to the findings obtained by *Balinski* in [4.79] which establishes criteria for the polydispersity of sodium silicates. To study the hardening kinetics *Jelinek* uses the ^{29}Si-NMR-spectroscopy and notes that the spectrum remains constant during the curing by ester. He infers from this that the well known Ilersche model of the sol gel transition does not apply here. As the major process of transformation of the sol to the gel, the aggregation of the particles in the hydrosol solution must be adopted accordingly. *Jelinek* theorizes that the production technology of water glass binders has an influence on the achievable strength. Two different binder solutions are investigated, one of which was produced using the conventional melting technology and the other through direct production of quartz sand and sodium hydroxide solution.

Herecova and *Jelinek* deal extensively with the applicable ester hardeners that are applicable in water glass ester processes [4.80]. Here it is stated that with the use of esters of acetic acid, the formation of insoluble sodium acetate is to be expected. This

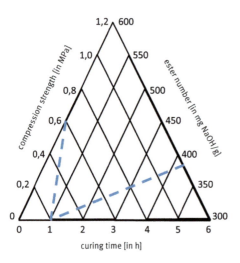

Figure 4.42: Nomograph to the correlation between ester number, sand strength and curing time [4.77]

sodium is responsible for a number of negative characteristics of the moulding material. For example, the thermoplastic behaviour of the moulding material and the diminished recoverability of used sands. One way to address some of these emerging issues is the use of esters of carbonic acid as hardener. One of these applicable hardeners is propylene carbonate through which the reaction, and thus curing time, can be significantly reduced. When this curing agent concentrate is used the solidification takes place very rapidly, and so this application is used only in exceptional cases. The authors examine various mixtures of propylene carbonate and the acetic acid ester, triacetin. Figure 4.43 shows that by mixing the two components it is possible to control the curing and thus the processing times of the moulding mixtures within wide limits. Figure 4.44 of the same publication shows that there is also a similar correlation from this to the residual strength and thus the collabsibility behaviour. When a mixture of propylene carbonate and triacetin is used, lower residual strength is expected up to temperatures of 800 °C as in the use of a mixture of triacetin, diacetin.

Studies on the effect of water glass modulus to the expected properties are done by *Pavlowsky* inter alia, in [4.81]. Here it is noted that both unmodified and modified binders systems are in use today. Possible modifiers such as silanes or boron are named. Among other curing technologies the water glass ester method is examined for the kinetics inherent in the curing. In order to make this curing visible, at least during the first phase of the reaction, the curing agent of triacetin (7 parts) and diacetin (3 parts) 5 % glycerolacetat was added. Besides the condensation through the elimination of water, a coagulation process during solidification was found. It can be shown through regular readings of the ^{29}Si-NMR-spectra that no clear distinction can be found between condensation and coagulation. Despite a significant change in the viscosity, the intensity ratios do not change significantly in the NMR spectra. One can nevertheless make certain statements as is made clear in figure 4.45. Cure time decreases the intensities of Q^0 and Q^2_6, while the content of Q^4 increases slightly. Practically, the gel

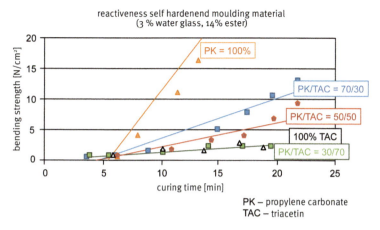

Figure 4.43: Response times of water glass ester moulding materials with hardening mixtures of propylene carbonate and triacetin [4.80]

ultimately produced approximately the same Q-type distribution as the output binder solution. With the ^{29}Si NMR spectroscopy, it is therefore possible to visualize the solidification sequence during the water glass ester method. In [4.82] *Glaß* talks about his experiences with the water glass ester method for the production of iron castings and is again discussing such important benefits such as job-friendliness, technological comparability with organic binder systems, and very low emission development. In his investigations *Glaß* uses water glass binder solutions with a modulus from 2.5 to 2.8; the hardeners are mixed from diacetin and triacetin in different proportions. Glycerine is used as a disintegrating component. In the representation of the used mould mate-

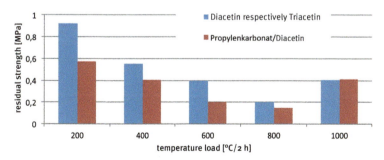

Figure 4:44: High temperature strength behaviour of water glass ester form of substances at different hardeners mixtures [4.80]

Cold self-curing processes | 103

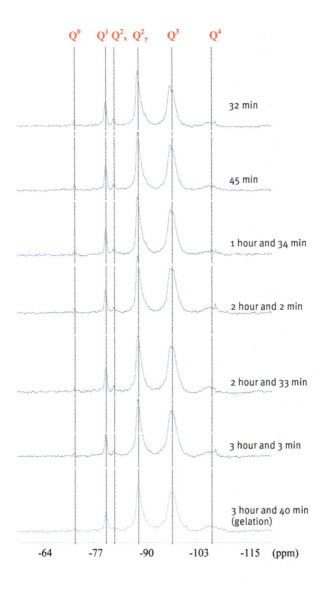

Figure 4.45: Sequence of solidification through water glass ester process, representation with the ^{29}Si NMR spectroscopy [4.81]

rial mixtures interesting formulas can be gleaned. The amount then, of the binder layer thickness (silica) sand grains 1 to 2.5 · 10^{-3} mm. From this it is derived that

% binder content multiplied by average grain size = 1.

This means that the amount of binder required is inversely proportional to the average grain size. The required amount of hardener is specified here based on 10 % of the binder quantity, the decay conveyor glycerin is added to the mould material mix at 0.06 %. The proportion of disintegrant also has an effect on the reaction rate as shown in figures 4.46 and 4.47. While figure 4.46 shows the pressure curves and shear strength for a mould mixture without glycerol, figure 4.47 displays the corresponding profiles when using higher glycerol content. In addition to the faster increase in strength it seems likely that higher final strengths can be expected. Significant successes in terms of the residual strength can be achieved by the very minimal addition of an organic decay conveyor as shown in fig 4.48. Unfortunately, it is not apparent from the cited source whether it is a mould material with glycerine in sample 1 or not. But even if the curves 1 and 2 operate with glycerine added, it must be taken into account the very low additive content. Also interesting are the conclusions of the author who reports out of his past professional practice. He notes that the water glass ester method allows a wide scope for the application of this technology both in terms of the achievable strength and processing characteristics. Particularly important is the finding that the procedure does not cause any significant problems in terms of the decay behaviour in the production of castings made from cast iron.

In the area of iron casting in Germany this process is unfortunately in the past. As discussed earlier, it has been used successfully in a number of aluminium foundries. An application in the casting temperature between aluminium and iron is copper casting. *Hemmann* inter alia deals with the application of the water glass ester process for the production of cast copper castings in [4.83]. The process here is applied to the manufacture of wings for controllable pitch propeller. The moulding box dimensions are varied within the limits of approximately 2.70 m x 2.20 m x 0.50 m and 4.40 m x 3.20 m x 0.7 m. Due to the specifics in this mould production (large flat shape mould boxes) a binder content of 3.5 % is used. A report on the development work aimed at improving the technological characteristics of this process is in [4.82]. Here again, among other properties, the bending strength and the disintegration behaviour are discussed (in this case with reference to the residual compression strength). Figures 4.49 and 4.50 show the development findings. The curves show very clearly that it is possible to achieve very impressive improvements through targeted chemical changes in the binder system.

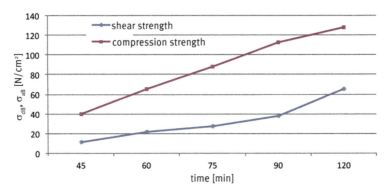

Figure 4.46: Compression and shear strengths without glycerol added [4.82]
(σ_{ab} – shear strength, σ_{db} – compression strength)

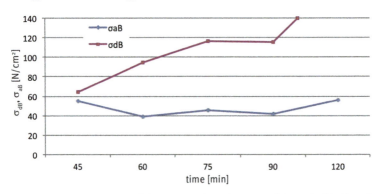

Figure 4:47: Compression and shear strengths with increased glycerol added [4.82]
(σ_{ab} – shear strength, σ_{db} – compression strength)

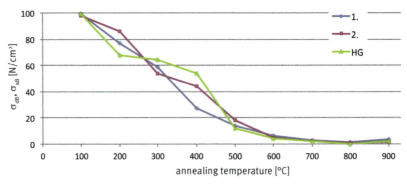

Figure 4.48: Residual shear strengths with and without glycerol added [4.82]
(σ_{ab} curve 1, σ_{db} – curve 2, HG with glycerol added)

Summary water glass ester process

The water glass ester process is currently the most promising cold self curing moulding process based on an inorganic binder. In principle, one can produce self curing moulds and cores with the same water glass binder system that is used for cores produced, for example, according to the water glass CO_2 process. However, because of (larger) hand moulds and the massive cores and other requirements that go with series cores, other binders systems should be used for these purposes. The strengths achievable are 2–4 times higher than that of carbon dioxide fumigated mouldings (here the 24-hour strength since the immediate strength is much lower). Strength-enhancing additives are therefore now largely unnecessary. Another reason why such additives in the binder can be dispensed with is that the flowability requirements here are not usually as high as in the area of core production. Decay behaviour in larger moulds is often of greater importance. These, and other requirements agreed upon today, show that with the water glass ester process, the binder can be used with little or no modification in as much as the added modifiers can be either organic or inorganic in nature. The method is currently used primarily in the manufacture of cast aluminium and cast copper. It has been displaced from the area of iron and steel casting, however this has essentially no technological reason. The main disadvantage is undoubtedly the fact that the recycling of the used sands is difficult. The well-known method from the cold resin reprocessing technology, in which the used sand is often just broken and dusted off, is not successful in water glass ester used sands. Here an increased mechanical energy is required to separate the old binder constituents from the sand grains. The subject of reclamation will be discussed in a later section. Future development studies must

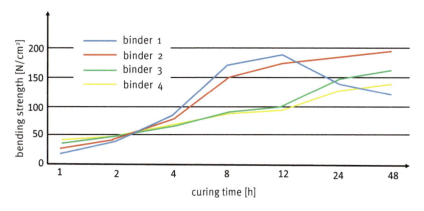

Figure 4.49: Development of bending strength with the water glass ester process by binder optimization (binder 1 – initial state) [4.83]

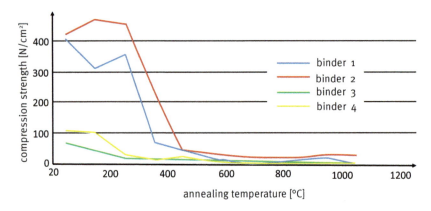

Figure 4.50: Evolution of the residual compression strength of water glass ester method by optimizing binder (binder 1 – initial state) [4.83]

focus on reclamation technologies. Under certain conditions and with the use of specific foundry materials the decay behaviour makes improvements of these properties necessary. In the field of cast alloy castings with high melting points such as cast steel, the thermoplastic behaviour brings occasional problems so this is also a point for further development. By addressing these points, water glass ester method may again increasingly be applied in the near future. If in the future it would be possible to develop an environmentally friendly and inorganic hardener, this method would be an ideal cold self curing moulding process.

4.2.3 Geopolymer process

Besides the water glass ester process, there has been another moulding technology in use called the geopolymer binder system which has been utilized for some years now. Here, the water glass binder related products are silicon-oxo-aluminate (see also chapter 3). The strength of the moulding material is formed by chains of SiO_4 and AlO_4 tetrahedra. Through the ratio of Si : Al it is possible to change the properties of the system. Binders usually used today operate with a ratio of Si : Al of 10 : 1. Examples of trade names are SIAL, Rudal or GeoPol. The geopolymer binder first emerged in the 90s in Poland and the Czech Republic, there are now some users in Germany, mainly in the aluminium casting industry. In addition to the self-hardening variants, this binder can also be used for curing with carbon dioxide. The self-hardening system consists of two components: the binder (an alkaline geopolymer) and the curing agent (a mixture of

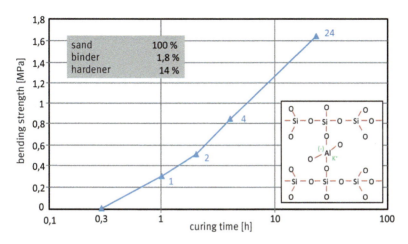

Figure 4.51: Typical characteristics geopolymer curing process
(18 min processability, stripping time 34 min) [4.88]

organic esters). Basically, the same curing agents can be used as in the water glass ester process, for example, esters of acetic acid or propylene carbonate. In order to control the hardening and the influence of the hardening characteristics it is possible to accelerate or decelerate the solidification process by the mixture of two hardening components. Thus, it is feasible to adapt the production process to changing environmental conditions. According to *Burian* inter alia [4.84], this method results in an almost linear increase of strength in the production of moulds (figure 4.51). The binder used has a designation of a 10:1 silicon and aluminium ratio. The binder contents are often shown in different published sources as being 2 to 3 %, however figure 4.51 shows the method also works well with binder content of less than 2 %. The curing

Figure 4.52: Water glass ester mould material – Cohesive fracture behaviour (binder content of 2.5 %) [4.88]

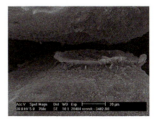

Figure 4.53: Geoploymer mould material – Adhesive fracture behaviour (binder content of 1.8 %) [4.88]

Figure 4.54: Moulds made of geopolymer bonded moulding material [4.86]

agent works analogous to the water glass ester process with a content of about 8–15 % based on the amount of binder used. As a major difference to the water glass process, *Burian* in [4.84] describes the tendency in geopolymers for adhesive fracture behaviour (figures 4.52 and 4.53). From this he concludes that the recycling of the used sand is possible and even easier to implement than with the water glass ester process. The reclamation is therefore similar to the furane resin process in that simple vibration crushing can be used. Therefore, in the manufacture of cores, 75 % of the regrind can be used while 100% reclaim is possible for backfill material.

Novotny and others give a number of examples of application of the process in the production of aluminium, copper, iron and steel castings [4.85]. When hardening by carbon dioxide gasification, such as in core production, a higher geopolymer binder content of between 2.5 and 3 % should be used. In [4.86] an overview of inorganic binder systems is shown including the geopolymer process. An example of a typical mould made from this form material is shown in figure 4.54. It is clear that mainly moulds and cores of simple geometries are made from this process.

Krahula and *Novotny* introduce the latest application status of the method in [4.87]. The development is accordingly expressed by the change in the name 'Rudal A' in

Geopol. Within these binder families the systems for the manufacturing of carbon dioxide gassed cores, the production of self hardened cores, and moulds such as large moulds are distinguished. An example of the production of steel castings with a maximum mass of 5 tons and a binder content of 2 % and a hardener content based on 15 % (relative to binder content) is given. The moulding material can be reprocessed in a mechanically operated recycling plant. A mixture of 75 % recycled and 25 % new sand is used for the applied moulding material and then backfilled with 100 % regenerate.

Literature – Chapter 4.2

[4.48] Nürnberg, M., Technologische Untersuchungen an alternativen Formgrundstoffen unter Verwendung anorganischer Bindersysteme, Studienarbeit, TU Bergakademie Freiberg, 2009
[4.49] Beilhack, M., Neue Erfahrungen beim Zementsand-Formverfahren, GIESSEREI 1952, Nr. 17, S. 405–410
[4.50] Endell, K., Über synthetische Gießereisande mit Zement als Bindemittel, GIESSEREI, 1942, Nr. 21, S. 349–356
[4.51] Buschmann, H., Über ein in Deutschland noch nicht allgemein bekanntes kastenloses Formverfahren, GIESSEREI, 1932, Nr. 19, S. 485
[4.52] Paschke, M., Waymann, C., Über die Verwendung von Zementen als Bindemittel bei Sanden für Gießereizwecke, GIESSEREI, 1941, Nr. 6, S. 134–135
[4.53] Locher, F. W., Zement – Grundlagen der Herstellung und Verwendung, Verlag Bau+Technik Düsseldorf, 2000
[4.54] Roll, F., Handbuch der Gießerei-Technik, Zweiter Band/2. Teil, Springer-Verlag 1963,
[4.55] Hemmann, U., Regenerierungsmethoden für Gießerei-Altsande/Praktische Anwendung in der MMG Mecklenburger Metallguß GmbH Waren
[4.56] Hemmann, U., Händler, E., Polzin, H., Der Betrieb einer Handformgießerei für Kupfergusslegierungen nach dem Wasserglas-Ester- und dem Zementsandformverfahren. GIESSEREI, Band 93 (2006) Heft 2, S. 38–43
[4.57] Wrase, M., Untersuchungen zur Optimierung des Zementformverfahrens, Großer Beleg, TU Bergakademie Freiberg, 2010
[4.58] Verein Deutscher Zementwerke e. V.: Hydratation des Zements und Gefüge des Zementsteins, Internetveröffentlichung S. 109–133, http://www.vdz-online.de/fileadmin/gruppen/vdz/3LiteraturRecherche/KompendiumZementBeton/1-4_Hydratation.pdf
[4.59] Eberlein, J., Polzin, H., Entwicklungstendenzen bei der Herstellung großer Schiffspropeller aus Aluminiumbronze, GIESSEREI-PRAXIS 2011, Heft 4, S. 131–135
[4.60] Henning, O., Kühl, A., Oelschläger, A., Phillip, O.: Technologie der Bindebaustoffe, Eigenschaften – Rohstoffe – Anwendung, Band 1, VEB Verlag für Bauwesen Berlin, 1989
[4.61] Röhling, S., Eifert, H., Kaden, R., Betonbau – Planung und Ausführung, Verlag Bauwesen Berlin, 2000
[4.62] Lohmeyer, G., Beton-Technik, Handbuch für Planer und Konstrukteure, Betonverlag, 1989
[4.63] Marx, S., Untersuchungen zum Verfestigungsverhalten von Zementformstoffen, Studienarbeit, TU Bergakademie Freiberg, Gießerei-Institut, 2006
[4.64] Wezel, A., Verfestigungsbeschleunigung von Zementformstoffen, Studienarbeit, TU Bergakademie Freiberg, Gießerei-Institut, 2010

[4.65] Tomek, L., Rusin, K., Stachovec, I., Influence of cement moulding sands on the structure of castings, Giessereiforschnung – International Foundry Research, Nr. 3-2010, S. 16–25
[4.66] Schuster, S., Anwendung selbsthärtender Formstoffe in der DDR, Gießereitechnik 1976, Nr. 12, S. 399–401
[4.67] Borsuk, P. A., Vasileva, G. I., Analyse des gegenwärtigen Zustandes und der Entwicklungstendenzen der technologischen Prozesse für die Herstellung von Gießformen und Kerne, Gießereitechnik 1976, Nr. 12, S. 401–406
[4.68] Beckius, K., Die Verwendung von Wasserglas und Esterhärtern, besonders mit Olivinsand, Gießereitechnik 1976, Nr. 12, S. 407–409
[4.69] Schumann, H., Lindemann, R., Chemisierung der Prozesse zur Herstellung von Kernen und Formen, Gießereitechnik 1976, Nr. 5, S. 156–158
[4.70] Mai, R., Schumann, H., Einsatz und Anwendung des Gisacodur-Verfahrens, Gießereitechnik 1975, Nr. 4, S. 126–128
[4.71] Gerstmann, O.; Erfahrungen bei der Anwendung von selbsthärtenden Wasserglasformstoffen, Gießereitechnik, 1976, Nr. 4, S. 122–130
[4.72] Burian, A., Kristek, J., Neue Auffassungen zu den Wasserglasbindern bringen technische und ökonomische Vorteile, Vortrag, 46. Internationaler Gießereikongreß, 1979
[4.73] Jelinek, P., Studium der Eigenschaften des Bindemittelsystemes Wasserglas – flüssiger Härter und Wasser, Freiberger Forschungsheft B 209, VEB Deutscher Verlag für Grundstoffindustrie Leipzig, 1979
[4.74] Jelinek, P., Petrikova, R., Hähnel, U., Flemming, E., Möglichkeiten der Modifizierung von Wasserglaslösungen durch Magnetfeld- und Ultraschallbehandlung und praktische Erfahrungen bei magnetfeldbehandelten Wasserglaslösungen in der CSSR (Teil 2), Gießereitechnik 1982, Nr. 1, S. 21–25
[4.75] Svensson, I. L.; A Scanning Electron Microscope Investigation of the Breakdown of an Ester-Cured Sodium Silicate Binder; AFS International Cast Metals Journal, September 1982, S. 32–42
[4.76] Flemming, E., Polzin, H., Kooyers, T. J., Beitrag zum Einsatz verbesserter Formtechnologien auf der Basis von Alkali-Silikat-Binderlösungen, GIESSEREI-PRAXIS 1996, Nr. 9/10, S. 177–183
[4.77] Sacharuk, L., Makarevitsh, A., Bast, J., Döpp, R., Bewertung der Reaktionsfähigkeit flüssiger Härter der Esterklasse mit Hilfe ihrer Esterzahl; Giessereiforschung 53 (2001), Nr. 3, S. 104–109
[4.78] Jelinek, P., Polzin, H., Strukturuntersuchungen und Festigkeitseigenschaften von Natrium-Silikat-Bindern, GIESSEREI-PRAXIS, 2003, Nr. 2, S. 51–60
[4.79] Balinski, A., Wybrane zagaddnienia technologii mas formerskich ze spojicznymi, Institut Odlevictwa, Krakow, 2000
[4.80] Herecova, L., Jelinek, P., Esters of carbonic acid – hardeners of binders on the base of sodium silicates; Transactions of the Technical Universtity of Ostrava, Metallurgical Series, Nr. 1, 2006, S. 79–86
[4.81] Pavlowsky, J., Thomas, B., Brendler, E., Polzin, H., Tilch, W., Skuta, R., Jelinek, P., Wasserglasgebundene Formstoffe? – Untersuchungen zum Einfluss des Silikatmoduls und der Verdünnung auf die Struktur und die Festigkeitseigenschaften von Wasserglaslösungen und zur Kinetik des Härtungsprozesses, GIESSEREI-PRAXIS 2005, Nr. 3, S. 1–7
[4.82] Glaß, W., Wasserglas-Ester-Formstoff für Gussstücke aus Gusseisen, GIESSEREI-PRAXIS, 2006, Nr. 1–2, S. 18–22

[4.83] Hemmann, U., Händler, E., Polzin, H., Der Betrieb einer Handformgießerei für Kupfergusslegierungen nach dem Wasserglas-Ester- und dem Zementsandformverfahren, GIESSEREI 93 – 2006, Nr. 2, S. 38–43
[4.84] Burian, A., Antoš, P. u. a., Kalthärtendes, anorganisches Formherstellungsverfahren auf Geopolymerbasis, Vortrag Sitzung WFO-Kommission Anorganische Binder am 18. April 2005, Milovy, Tschechische Republik
[4.85] Novotny, J., u. a., Das anorganische Bindersystem Rudal/Geopol, Vortrag Sitzung WFO-Kommission Anorganische Binder am 21. April 2008, Milovy, Tschechische Republik
[4.86] Polzin, H., Die Anwendung kaltselbsthärtender anorganischer Bindersysteme, GIESSEREI-PRAXIS 2010, Nr. 9, S. 282–287
[4.87] Krahula, Z., Novotny, J., Das anorganische Bindersystem Geopol und seine Einführung in die Gießereien, Vortrag DBU-Tagung Osnabrück, 2010

4.3 Warm or hot curing processes

4.3.1 Processses with tempered primary tools

The solidification of water glass bonded moulding materials in heated primary tools, core boxes, etc., has long been known. Certainly on occasion, inorganic (water glass) cores were produced in the existing metallic tools for core production in the heyday of the hot box process. Although the achievable strength could compete with those of the hot box process, the "water glass hot box process" has not become established. In addition to the high energy required for manufacturing considerations, collapsibility is likely to have played an important role in this. With the resurgence of interest in the inorganic binder systems, the heat-curing process variants are again being explored and are the focus of some discussion.

In [4.88] *Svensson* deals with the CO_2 hardening through water glass binders, but also notes that the drying brings good results after the hot box process with addition of carbon dioxide gassing. *Svensson* points to good finishes, better decomposition,and better storage stability in humid environments. The interior of the core thus cures more quickly than with pure warm air gassing. The partial pressure of carbon dioxide influences the type and amount of the forming silicate species. Calculations show the distribution of silica in the binder solutions with an increasing pressure and elevated concentration of CO_2. About 1–2 mol of carbon dioxide can therefore be dissolved per liter of binder before forming sodium carbonate. With higher CO_2 concentrations in the fumigation (or drying), more carbonates are formed. *Osterberg* and *Anderson* reported in [4.89] on the Saab-Scania process in which a water glass sand mix were shot in 130–150 °C hot core boxes and then gassed with 120 seconds of 170–190 °C warm air, or alternately with warm air and carbon dioxide. The process resulted in good strengths with a 2.4 to 3.2 % addition of binder. Also, the mould material was sufficiently fluid and showed good disintegration behaviour. In other research *Flemming* discovered inter alia in [4.90] that with regards to the most complete utilization of the binding properties of water glass binders the best results are achieved during dehydration due to the formation of compact binder bridges. It is also pointed out that when using modified sodium water glasses, the use of a hot-box technology with binder contents of 3 % is advantageous. In [4.91] and [4.92] *Döpp, Schneider* and others carry out tests in which they shoot the moulding material into warm core boxes. The cores are prepared at room temperature as compared with those prepared in 50, 100, 150 and 200 °C warm core boxes. After the shooting, the authors carried out additional gassings of carbon dioxide with binder levels at 2, 2.5 and 3 %, a method with which cores for motorcycle cylinder heads are produced. However, with a binder content of 2 %, sufficient strength cannot be obtained. *Doroshenko* and *Makarevich* also work with hot core boxes in [4.93]. Here they work with 2 % water glass moulded masks and cores that have a tensile strength of more than 2 MPa. They use a binder modified with alkali salts, salts of tripole phosphororic acid and silicon-organic compounds. The core box temperatures in these investigations are 180–240 °C.

Starting with the GIFA 2003, work on the development of new hot-curing Binder systems were introduced by different binder manufacturers. These developments are still influencing today's industry. The foundation of these developments was the known fact that water glass bonded moulding materials can be taken to significantly higher strengths through the use of tempered forming tools, for example, in the case of the classic carbon dioxide fumigation. These high strengths were found to be a fundamental requirement of inorganic binder systems since one of the main impetus for this development came from the automotive industry, and thus emanated from the large-scale production of some highly complex and delicate cores for vehicle components. In addition to the binder systems Inotec® and Cordis® systems based on modified silicate, or water glass binder solutions (which are discussed in the next section), the salt binder systems Hydro Bond® and Laempe Kuhs® emerged out of GIFA 2003 as two real new developments in the market. Although the various binder systems that are based on salt compounds have now been overtaken in importance by silicate binders, they are nonetheless discussed here.

Binder system Inotec® [4.95]
Inotec is a silicate-based binder. It is therefore related to the water glasses that are used in the water glass CO_2 process. However, unlike the known CO_2 method, curing does not take place on a purely chemical basis, but is based on a combination of physical drying process and chemical reaction. This is also the reason why heated core boxes are used during the Inotec® process. With the temperature of the core box at 150–200 °C the water serves as a solvent for the silicates and is then expelled from the core. There is also a chemical reaction that is initiated which leads to the cross-linking of silica molecules. The byproduct of this connection is also water (chemically there is a condensation reaction, meaning a reaction in which two molecules combine while splitting off a smaller molecule, in this case water). Through this, liquid from two different sources (solvent and reaction byproduct) is freed. Ideally, this chemical-physical curing process is assisted by a hot air purge (ca. 150 °C) in order to reduce cycle times during the core production through a maximum expulsion of moisture from the core. The curing reaction is partially reversible (equilibrium reaction), that is to say with the input of energy and water (for example high temperature and humidity) the reverse

```
Inotec process
    │
    ▼
┌─────────────┐  silicate binder mixture
│ preparation │  2-3 % + promotors:
└─────────────┘  0,2-0,6%
    │
    ▼
┌─────────────┐
│ compaction  │  p=0,4-0,5 MPa
└─────────────┘
    │
    ▼
┌─────────────┐  T_H = 160-200°C
│thermic curing│ σ_b = 200-300N/cm²
│ (warm box)  │  shell evolution
└─────────────┘
    │
    ▼
┌─────────────┐
│  purging    │
│  hot air    │──► full hardening
│ or cold air │
└─────────────┘
    │
    ▼
┌─────────────┐
│stripping of core│
│ core after  │
│ treatment   │
└─────────────┘
```

T_H = 160-200°C
σ_b = 200-300 N/cm²

Figure 4.55: Inotec® Binder system procedure

reaction can take place whereby the network of silicates is removed resulting in the core losing strength and breaking. This can be prevented by removing water from this balance through dry storage. In practice, this is not always so easy so additives (called promoters) are used. This delay in reverse reaction significantly improves the safe handling of the cores even after normal storage and is referred to by *Müller et al.* [4.96, 4.97] and *Wallenhorst et al.* [4.98, 4.99].

The advantages of using this inorganic binder system are not limited solely to the reduction of odors and emissions (*Wallenhorst et al.* [4.99], *Weissenbeck* et al. [4.100]). It also results in significant benefits in cleaning and maintenance of the casting tools. Moulds, which had to be blasted at relatively short intervals before because they were contaminated with organic condensates, can be sometimes operated around the clock. This is not only a cost-reducing effect, but increases productivity significantly. By the absence of condensate build-up in the mould, the temperatures in the base plate portion can be reduced, resulting in a faster solidification. The advantages are obvious: on the one hand, shorter cycle times, (productivity gain), on the other hand a shortened dendrite arm spacing through rapid cooling (*Pabel* et al. [4.101]). The latter directly affects the static and dynamic properties of the compo-

Figure 4.56: Inotec® core, crankcase (image BMW AG, ASK Chemicals)

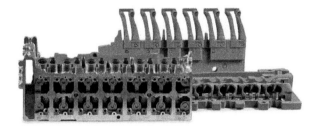

Figure 4.57: Inotec® core, cylinder head (image BMW AG, ASK Chemicals)

nents and can be seen as further technological progress toward downsizing. In the construction of new low-emission but powerful engines, these improved structural properties can be considered during the designing of casting wall thickness (*Kautz et al.* [4.102], *Beck* and others [4.103]).

Comprehensive studies on the cycle behaviour of the used inorganic core sands have shown that reclamation in high proportions (> 90 %) is possible (*Schwickal et al.* [4.104]). Series production of the first commercial plant has been made and confirms the good results of the test series. The transfer of the inorganic binder approach to iron casting has also already taken place (*Sasse et al.* [4.105]). The next few years will show whether here, as in aluminium permanent mould castings, other benefits are to be expected along with the emission reduction. It is conceivable that, for example, casting defects such as gas bubbles or veins can be reduced through the use of inorganic binders or even eliminated.

Binder system Cordis® [4.106]

The binder system Cordis® is also used to make inorganically bound cores in heated core production tools [4.107]. The Cordis® binder system consists of water as a solvent, and an inorganic binder matrix. This binder matrix consists of a combination of modified phosphate, silicate and borate groups depending on the application. Adaption to specific applications is possible because a combination of salt groups. Furthermore, the characteristics can be specifically controlled by addition of inorganic substances directly into the binders or as an additive during production of the cores. These characteristics include, for example, the fluidity, the reactivity of the moulding material mixture, the wetting of the core by the melt, and the shelf life. The core production is carried out in commercial core shooting machines with the possibility of using heated core boxes. In the preparation of the core, moulding material is shot into the heated core box. The core box temperature is between 120 and 160 °C depending on the geometry; a homogeneous heat distribution in the core box is desirable. At the begin-

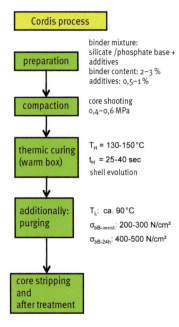

Figure 4.58: Cordis® binder system, procedure

ning of the core hardening, a shell forms along the outer contour of the core. This curing process is based purely on physical dehydration of the solvent. Depending on the type of Cordis® a possible additional chemical curing is possible, whereby higher extraction strengths are achievable. After cooling to room temperature, the core has reached its ultimate strength and can be poured off. Depending on the type of binder and sand-binder, additions are typically from 1.5 to 3 %. The cold bending strengths achievable are 350 to 550 N/cm². Investigations of bending strength based on standard tests showed that the strength values of the cold box in the Cordis® system are almost identical between 1 and 24 hours after preparation. Accelerated curing is possible by hot air fumigation whereby an improved warming of the mould material results in a more efficient removal of the released water. However, the curing time is generally dependent on the geometry and type of core boxes: The bigger and more compact the core, the more difficult it is to cure. The good flowability of the moulding mixture allows the production of cores with intricate contours which are necessary, for example, for retarder and channel cores, or cores for automotive parts such as the production of complete core sets for cylinder heads. When casting aluminium alloys, unfinished cores may be imple-

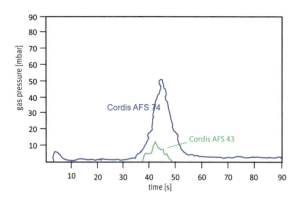

Figure 4.59: Gas pressure curves of two Cordis® cores with different sand grain sizes sizes [4.106]

mented. However, should a finishing be required, alcohol and water coatings can be used. The finished cores lose strength only slightly. However, the use of water coatings lead to higher strength losses than by using alcohol coatings. In the core storage, it should be noted, that the cores have a highly hydrophilic character, and thus should not be stored at high humidity. Twenty four hours storage at high relative humidity causes a loss of strength by a third.

During the casting phase, fume development is very low and an odour cannot be detected. Even after several casts, no condensation on the mould is detectable. In order to perform a quantitative analysis of the occurring additional pyrolysis products, a casting device was developed in a Hannover foundry which assisted in the determination of condensation and odour during casting of aluminium alloys. Unlike organic binder systems, no condensation within a measurement of accuracy could be detected with the Cordis® system. Organic systems showed a significant condensation, which is a common cause of casting defects. Just like organic systems, in the Cordis® system there is also air formed in the pore spaces during pouring but with different characteristics. Because of the non-uniform geometry of the grains of sand, many voids remain in the core. By using Archimedes' principle, pore volume can be determined. Investigations revealed pore volume of more than 30 %. Thus, a relationship between the grain size and quantitative gas pressure can be detected. The use of a coarse sand gives a significantly lower gas surge than by the use of a fine-grained sand (fig. 4.59). While the organic cores immediately and continuously show gases from the time of melt, the Cordis® cores show a significant gas pressure only after 30 to 40 seconds after melt contact whereby a maximum is reached relatively quickly. This is followed by an immediate drop in the gas pressure (fig. 4.60), *Voigt et al.* [4.108], *Löchte et al.* [4.109]. The Cordis® system achieved good results in the aluminium gravity die casting, green sand

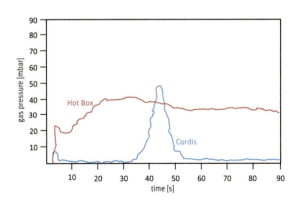

Figure 4.60: Gas pressure curves of hot box and Cordis mould materials [4.106]

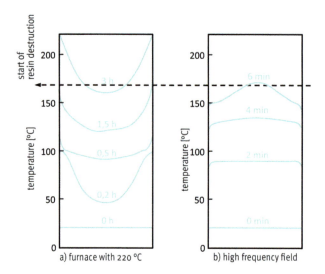

Figure 4.61:
Temperature profile in a cylindrical core when heated by thermal conduction and by high-frequency energy [4.111]

cast aluminium, and aluminium low pressure casting. The surface quality compared to organic systems is similar or improved. Core removal of the castings can be performed on conventional decoring devices without prior heat treatment. The secondary strengths are higher compared with those in the cold-box system because organic binder systems are pyrolized under thermal stress while with the Cordis® binder system an embrittlement of the binder matrix must take place after thermal gassing. This has the effect of collapsing the binder bridges and consequently causes dissolution of the polymeric network according to *Voigt et al.* [4.108]. In early 2005 pilot tests were started in series production at the Hannover Volkswagen foundry where the previous findings along with the validated laboratory results were incorporated into the development. The first experiments were carried out on intake pipes and two cylinder head types. The production of the Cordis® cores is done on modified series tools. The modifications are done according to the different temperature ranges the tools are used in, the option of heating the core box, the integration of a heated gas system, and optimization of the core box vents in the core engraving. Due to the very different flow characteristics in the Cordis® system and organic systems, an adaptation of the shot image of the tools is also required. Curing is carried out through a combination of hardening in the warm mould at 140 °C and additional hot air purging with temperatures of approx. 90 °C. The initial strengths correspond to those of the previously used organic systems, with the binder quantities specified at 2 to 3 %. The curing times are between 25 and 40 seconds. Because of the selective moistening of the slot nozzles, they need not be cleaned and a release agent is not necessary. The casting of the core packages is carried out in the gravity casting where the shells, the casting parameters, the melting temperature, the alloy, as well as the cycle time, correspond to the previous series production conditions using the organic binder. This also applies to the handling of the cores by the caster.

During the casting, no contamination by condensation is observable. Comprehensive analysis of quality, which included CT scans, endoscopic examinations (which included sections of the casting), showed neither penetration nor sand buildup. There were also no bubbles or other casting defects across a full dimensional line of castings. Cordis® cores have been successfully used in series production for some years. Literature on the subject of reclamation of Cordis® used sands or recycling plants are not found according to *Voigt et al.* [4.108] and *Bischoff et al.* [4.110].

4.3.2 Microwave drying method

Foundational and early developments
It is well known that the physical hardening of water glass solutions by drying provides strengths in higher ranges than that with other hardening technologies such as carbon dioxide, ester hardening, etc. The acceleration of drying by the use of microwave energy for the curing of shaped parts made of silicate-bound moulding material systems is a way to make better use of the binding potential of these systems. First investigations of microwave drying water glass bonded mould materials were already carried out in the 1960s. For example, *Grassmann* worked with high frequency drying as a way to cure water glass bonded mould material [4.111]. In his investigations he used a capacitor field from two metal capacitor plates through which the high-frequency voltage of a tube generator is run. *Grassmann* emphasizes that as a special feature of the dielectric heating, a core can be heated evenly across an entire cross-section within minutes. The process of convection drying works with drying occurring from the outside to the inside and takes much longer. The two contrasting heating mechanisms are shown clearly in figure 4.61.

For these tests, core sand mixtures were used containing 1 to 6 % water glass binder. Carbon dioxide pre-hardening and high frequency energy were used exclusively. The results showed that for high-frequency drying, water-glass binder having a modulus from 2.3 to 2.5 are more suitable because they do not react as fast with the carbon dioxide in the air. By such reactions the strength of the core surface, and thus the edge strength and abrasion resistance, are reduced. In summary it can be said that through the uniform drying of cores by high frequency energy, good results can be achieved in

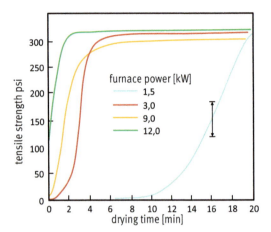

Figure 4.62: Tensile strength with increasing drying time according to microwave power [4.114]

terms of core strength, dimensional stability and surface quality. The pre-curing with carbon dioxide to achieve a green strength of the cores affects the achievable (by drying) ultimate strength considerably.

In [4.112] *Schreyer* also discusses inter alia the advantages of the evenly distributed heating achievable by high frequency drying in core production of water glass bound mould material mixtures. It is noted that the transmission capacity for the heat output in the high-frequency field is approximately 20,000 times greater than with convection drying. *Tripsa* and others describe in [4.113] tests for core drying with a 1 kW unit in which they use both water glass binder and binder for the ceramic moulding process. They also note significantly higher strengths achieved at approx 1/10 of the curing time compared to conventional curing technologies. The use of carbon dioxide to pre-cure is judged as negative. According to these authors, microwave drying is a way to considerably increase the productivity in the core production of water glass bound mould materials due to the automation of the process. *Cole* also deals inter alia with the microwave drying of silicate bonded cores [4.114]. For their investigations they use four different water glass binders with modules between 2.0 and 3.22 and differing water contents. The microwave oven used had a power of 5 kW. The effects of additives such as zinc oxide, chromium oxide, starch, coal dust, synthetic resin, cupola furnace slag, iron oxide, and phosphoric acid on the forming properties were examined. Zinc oxide and starch additives are found to reduce the strength loss during storage of cores in a moist environment while improving tensile strength and flexibility. By the addition of chromium oxide, moisture sensitivity is reduced. Without addition of these modifiers, the moisture sensitivity of the moulded parts cured by microwave energy is

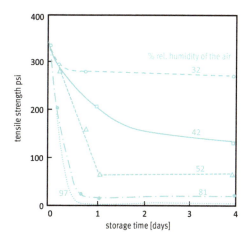

Figure 4.63: Tensile strength during storage at different relative humidities [4.114]

very high. With the modifiers added, the cores can be stored in 80 % atmospheric humidity. Figure 4.63 shows the tensile strength correlating to the core storage time at different relative humidity.

Owusu and *Draper* investigate ways of improving the moisture resistance of water glass bonded mould materials in microwave hardening [4.115] and [4.116]. To achieve a better resistance to moisture in the moulding mixtures, magnesium, lithium, calcium or zinc carbonate is added. The addition of lithium carbonate or silicate gives the best results, achieving over 90 % of initial strength after 24 hours storage in 97 % relative humidity. Such additives also improve the unpacking behaviour considerably. The presence of sodium ions and hydroxyl groups is at first and foremost responsible for the hygroscopicity. Na_2O and SiO_2 form eutectic compounds having low melting points which can lead to difficulty in the disintegration of the core. By the use of additives, property improvements can be achieved. The microwave oven used for the investigations has a power output of 6 kW, the curing time is 5 minutes. The authors come to the conclusion that the highest strengths are obtained when curing water glass mould materials by hot air. Much shorter microwave curing times results in only slightly lower strengths. In further publications from *Owusu* and *Draper* the moisture resistance of water glass bonded materials during microwave drying is the subject of their investigations [4.117]. It is found that the loss of strength is reduced with increasing proportion of $SiO_2 : Na_2O$. "Normal" water glass binders are quite unstable at a humidity of over 50 %. However, if silicate binders modified with chromium or zirconium are used, cores hardened in a microwave oven can withstand a relative humidity of 80 %. The core strength is therefore defined as a function of silicate modulus, which is decisive

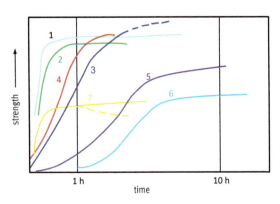

Figure 4.64: Strength profile of water glass materials with different bonding technologies [3.7]

for the water absorption of the system. An important finding in this work is that the strength can be fully restored in cores "damaged" by moisture absorption through short and repeated microwave drying.

Ailin-Pyzik et al. come to the similar conclusion that by the addition of a lithium component, a water glass binder resistant to moisture can be made [4.118]. Lithium compounds result in a reduced solubility of the binder while microwave drying reduces the amount of binder required. This reduces problems in both the decay behaviour and the recycling of the cast moulding materials. Microwave drying is found to be very beneficial. In addition to the already mentioned advantages of lower binder content, improved decay, and easier recycling of used sands, there is also lower sodium content in the moulding material and a smaller reduction of initial sintering temperatures. The water glass binders used in these studies have a modulus in the area of 1.95 to 2.47. The microwave oven used has a power output of 1.2 kW. According to these authors, the difference between microwave drying and carbon dioxide gassing is that when drying by microwave energy, the gel phase is formed by dehydration before it can be formed as a result of acidification caused by carbon dioxide. Significantly higher strengths come through the much more homogeneous binder films caused during the drying process. An example of the practical application of drying water glass bonded materials by microwave is given by *Carlson* and *Thyberg* in [4.119] with reference to the production of cylinder blocks. They note that the use of microwave ovens is the best way to improve the properties of water glass mould materials. *Chang* and others refer to the improved thermal decomposition of microwave cured water glass cores in

[4.120]. They carry out investigations in clay-modified binder systems. The produced cores have a considerably higher (tensile) strength and therefore an improved decay behaviour. The authors give a binder content of 1.5 % and a clay content of 0.5 % as an optimal formulation. *Flemming* and *Mende* point out that water glass solutions with a modulus of less than 2.5 are more suitable for drying (i.e. also for microwave drying) than solutions with higher modulus [4.121].

In [4.122] the production of a block shape with 5 % water glass and cured with microwave energy is described. To achieve adequate strength manipulation of the compressed moulding material it is briefly gassed with carbon dioxide. The final strength is then obtained by drying with a microwave. The power of the furnace here is 0.7 kW. The unpacking of the produced moulding piece is done with water. *Flemming* and *Tilch* confirm that the microwave hardening of water glass binders yields the best strength values [3.7]. Figure 4.64 shows the expected strength curves for different hardening variants in comparison. *Jelinek* assumes in [4.123] that a bending strength of 3 MPa can be achieved in microwave drying with a binder content of 2 %. The water glass used here has a modulus of 2.3; good core decay is also observed. The block forms produced are then processed on DISAMATIC mould systems. A silicate binder, which is particularly suitable for the microwave drying through modification with Al_2O_3, is presented in [4.124]. Besides good binder decomposition it also has low moisture sensitivity. Likewise, *Xiao* et al. also include the water glass microwave method in their studies in [4.125]. They also emphasize the high strengths through which binder content is lowered and reclamation behaviour can be improved. They note further that through the development of microwave technology in recent years, there are good opportunities for the increased use of this method in the foundry industry. Scanning electron microscopy is used to determine the characteristics of the binder bridges. The size of the resulting silicate particles, which are obtained by microwave drying and carbon dioxide gassing, is also determined. With the addition of a binder of 2.5 %, the authors reach a tensile strength of 2.4 MPa, while only 0.4 MPa can be achieved in CO_2 method with 5 % binder. The cores produced through microwave drying show a decay which is comparable to resin binders. When looking at the resulting binder bridges, it is determined that smooth and homogeneous binder bridges at significantly smaller silicate particles exist through microwave hardening.

Döpp and others experiment with the method using a domestic microwave oven with a power of 0.8 kW and a frequency of 2450 MHz [4.126]. They examine the bending strength of specimens containing 4 % water glass binder. The drying time is between 30 seconds and 10 minutes. The authors establish water losses at 1.4 % after 1 to 2 minutes drying time which results in maximum strength. Compared here are the water glass CO_2 process, the microwave hardening method with or without water, and microwave hardening with vinegar and sugar addition. The authors come to the conclusion that microwave drying is the ideal method for hardening of water glass materials since 80 to 90 % of the water is removed. With this method one can reach 2 to 3 times higher strengths compared to carbon dioxide gassing and by the addition of vinegar or sugar, strengths of 4 to 5 times higher can be achieved. *Alekassir* confirms and complements these findings in [4.127]. Using binder content of between 2.5 % and 5.0 % *Piegel* and *Granat* perform studies with a 0.8 kW microwave system [4.128] and [4.129]. The binder used has a modulus of 2.8 and 0.5 % water content is added to the mould material mixture. The curing times in the microwave range from 60 to 360 seconds. At 60 seconds the authors find no measurable strength and at 120 seconds they determine a partial curing. At 180 seconds drying time the samples are completely cured. For binder content of 2.5 % a value of 930 N/cm^2 bending strength is measured after 240 seconds. Tensile and compressive strengths are also measured along with gas permeability. After 24 hours of storage of the samples, *Piegel* and *Granat* note a deterioration of properties by 50 %. No specific details are given to the environmental conditions.

Using existing knowledge and test results, the technological behaviour of currently available water glass binder systems for microwave drying is considered in detail in [4.130]. In addition to the usual foundry technological tests this publication also spans the arc of analytical study methods for clarification of structural changes in solidification technology to the practical testing of previously developed technology to the production of real castings. The water glass binders used in this work include water glass binder of different moduli, modified binder systems, as well as a potassium water glass. Strengthening by microwave energy is compared with (slower) oven curing technologies and carbon dioxide gassing or combinations of the two. Figure 4.65 shows the test program; the industrial microwave furnace used had a maximum power output of 4 kW.

It is also found here that through the microwave drying of water glass bonded mould materials, maximum strengths can be achieved that are 8 to 10 times higher than through carbon dioxide gassing. With increased drying times even higher strengths can be obtained due to the complete removal of water from the system after 90 to 120 seconds. It should be noted that for these test samples there were longer drying times due to the fact that the microwave oven used had a fairly large application area relative to the size of the sample. In actual core production, appropriately adapted microwave furnaces make shorter drying times possible. With regard to the strength behaviour it was observed that that the best results were achieved with unmodified water glasses

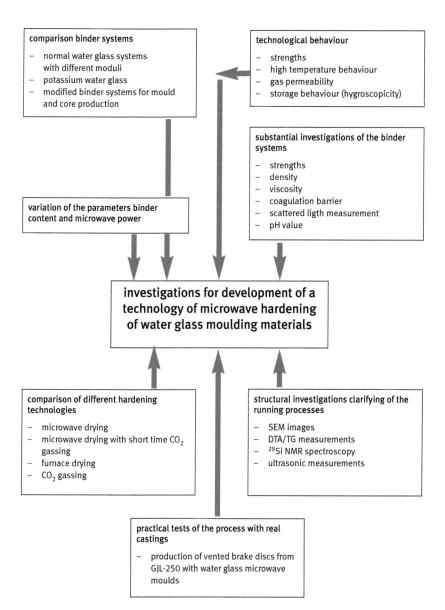

Figure 4.65: Targets and influencing factors of the studies conducted on the microwave hardening of water glass bonded mould systems [4.130]

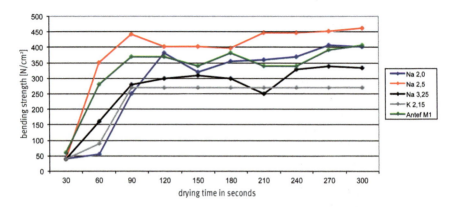

Figure 4.66: Bending strength by microwave drying and the comparison of different binder systems at 3 % binder content [4.130]

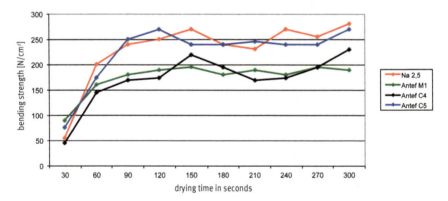

Figure 4.67: Bending strength of microwave drying with pre gassing of 5 s–CO_2, binder content of 3 % [4.130]

with a modulus under 3.0 (here 2.22 and 2.75). Sodium water glass (modulus 3.25) and potassium water glass (modulus 2.15) are less suitable for this hardening variant. Also tested were modified organic and inorganic modified binder systems both of which showed poor strength properties.

Figure 4.67 shows that a short-time (5s) gassing with carbon dioxide and subsequent microwave drying causes significant loss of strength. Pre-curing with carbon dioxide for the purpose of achieving faster manipulation-strength in the produced cores is not

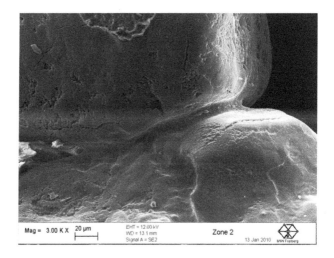

Figure 4.68: Water glass binder bridge after microwave drying [4.130]

advisable for microwave drying. Through furnace drying, strengths can be achieved comparable to microwave solidification but much longer drying times are needed. A problem in high-temperature behaviour was determined to be in the residual compression strengths. In both microwave and furnace drying a higher strength was reached than through the carbon dioxide method when done in a range of up to 200 °C. This is due to the significantly higher level of initial strength. In the range between 400 °C and 600 °C lower residual strength was found in comparison to the carbon dioxide variant. From about 600 °C no significant differences between the curing technologies were visible.

It is important to note that the storage stability of microwave dried cores is lower than that of cores cured by oven drying or carbon dioxide gassing. In any storage time, under normal conditions (which is assumed to be 70 % relative humidity) a storage time life of 24 hours is available, and in most cases even longer. The unmodified water glass binders with a modulus of 2.75 proved to have the best storage stability under these conditions or even at 90 % humidity, while the modified binder systems were prone to moisture. In summary, it is determined that when microwave drying a binder content of a 2 % maximum is appropriate in many applications except for in the case of simple core geometries where even lower levels are possible [4.130]. Of course, the required microwave power is dependent on the volume of the form parts to be produced. In the interest of short production times however, it should be set as high as possible. In the study of structure developments it has been confirmed that SEM images are suit-

Figure 4.69: Pre-hardened brake disc core in the core shooting machine, mass 1.7 kg [4.130]

Figure 4.70: Brake disc core after hardening in the microwave for 3 minutes, 3 kW [4.130]

able for the detection of differences in binder bridge morphology in the curing method variants. The homogeneous binder bridges that exist without interfering factors demonstrate the high strength measured at microwave drying (figure 4.68). Through the application of differential thermal analysis and thermogravimetry it can be demonstrated that the formation of strength-reducing carbonates happens rarely or only to a small extent in microwave drying. Investigations with the ^{29}Si-NMR showed that the unmodified water glass systems in the initial state have a more favorable degree of pre-condensation as that of the modified systems. However, this was only clear in binders with a modulus less than 3.0. It is further noted that organic modifiers obviously hinder the condensation of the systems. Ultrasonic measuring technology is also therefore suitable to visualize differences in the binding structures. Cores for ventilated brake discs were mentioned in the cited work to demonstrate the practicality of the water glass microwave method. The cores of brake discs out of GJL-250 had previously been made in the PUR-cold box process. The appropriate core tool had to be heated for the tests at temperatures between 100 and 120 °C. After 5 minutes, the finished core could be removed from the core box and transferred to a microwave oven for hardening. Parallel test specimens produced had a bending strength of more than 400 N/cm². In conjunction with a block form of water glass ester moulding material the production of a quality-oriented disc is possible. Difficulties in coring the thin webs inside the disc are solved by lower binder additions or targeted mixtures of additives. The same applies to the cycle times that are equally optimized with core manufacturing tools specifically tailored to the process. The results presented here are excerpts published in [4.131] and [4.132].

AWB® process

The AWB® process according to *Wolff* and *Steinhäuser* [4.133] is based on the thermal curing of water glass bonded mould material with a temperature controlled tool followed by microwave drying. The binder is a modified water glass, which has a low viscosity through dilution with sodium hydroxide. The flowability of the moulding mixtures thus produced, and thus their shootability, are only slightly less than that of PUR-cold box moulding materials. The solidification of the moulding material in the AWB® process takes place exclusively via dehydration at a mould temperature between 160 and 200 °C, whereby in addition, a vacuum can be applied. The hardening times are given as 10 to 60 seconds. The final drying is then ensured by a low power microwave oven at a time of several minutes. The binder additions lay between 1.5 and 2.5 % and no additives are used. Comparisons of gas development with the PUR-cold-box method as expected fall very clearly in favor of the AWB® process. In [4.134], *Gosch* points to good disintegration as a major advantage. According to *Steinhäuser et al.* [4.135], rough surfaces can be avoided though good wetability of the mould material by using alternative mold materials based on Al_2O_3 or MgO. Molten aluminium, for example, is a difficult casting material in regard to wetability. A further and significantly less expensive variant against the deterioration of wetting behaviour is the addition of additives which, through the formation of the Lotus Flower effect (on a particularly rough surface), prevent larger form material buildup on the surface of the casting part.

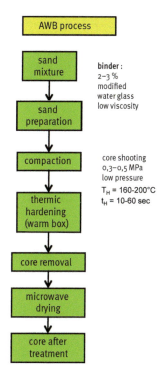

Figure 4.71:
AWB® process flow schematic

Summary of microwave drying

The hardening of water glass bonded moulding materials by microwave drying is technically a very interesting method. Due to the combination of short cycle times and environmental friendliness, microwave drying is a high performance core manufacturing process. Due to the high achievable strengths, the binder content can be set at 2 % and probably even lower which has a positive influence on disintegration behaviours. Although the industrial microwave technology has made great progress in recent years,

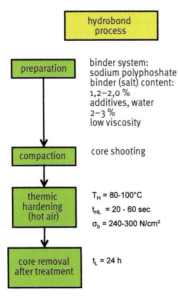

Figure 4.72: Schematic diagram of the Hydrobond® process

the stoves are still quite expensive and require some care and maintenance for daily use. However, both the technique and the microwave equipment has proven to be very "foundry-compatible" in a number of foundries that use them for drying and coating. The biggest challenge of this technology continues to be the core box material. The necessary microwave-transparent work materials such as teflon, quartz glass, or concrete are not particularly wear resistant or malleable. When the solution of the problem is successful and one follows the path to core production that is shown in green in figure 3.13, a breakthrough for this interesting method is achieved. A second very useful step forward would be the integration of microwave technology into the actual core production system, since an external drying unit would then be unnecessary.

4.3.3 Processes with salt binder systems

The real discoveries made in recent years in the field of inorganic core binder systems (and other binder systems) have been through the use of sand as moulding material. Generally this has been silica sand. In the case of this method the binder materials used are water soluble salts or salt compounds. The solvent water used ensures the necessary wetability of the grains of sand with the binder. The solidification of the cores is achieved through dehydration in a heated form tool. *Sobczyk* gives an overview of the following binder systems in [4.136].

Hydrobond® process

Under the name HydroBond®, new binder systems and system solutions were put on the market for core production at the GIFA 2003. The binders have been designed mainly for non-ferrous casting. The binder system is a water-soluble, inorganic system based on sodium polyphosphate which is solidified by the removal of water in the mould. The main components of the binder are therefore salt compounds, water solvent, and a number of specific additives which are primarily inorganic in nature. In core production, conventional core shooting machines are used wherein after the shooting a pressurized hot warm fumigation of the cores is carried out at temperatures of 80 °C. Through temperature increase, the salt compounds crystallize and strength is produced in the core; *Tilch* [4.137] and *Hänsel* [4.138]. The procedure is shown schematically in figure 4.72.

According to the referenced publications, the bending strength attainable with this method is a maximum of 250 N/cm² with a binder content of 1.5 % by mass. However, this can be slightly lowered through longer storage times of the mixed substances due to drying phenomena. As a special promotional aspect it was communicated at the time of its introduction that the binder had an influence on food quality. Since harmful emissions naturally occur in this process only to a very limited extent, problems with core decay after casting or the occurrence of specific defects could

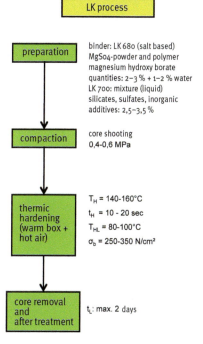

Figure 4.73: Schematic diagram of the LK® process

therefore not be observed. Through the recycling of the used sands one comes to the topic of wet reclamation which has already been suggested here by the reference to wet decoring. According to *Hänsel* in [4.138] up to 85 % of the water can be re-used in the re-processing of the material. Through phosphate recovery the binder components can be recovered and applied as secondary raw material in other recycling efforts. Maximum reclaimed sand usage is given in the cited references as 100 %. This was indeed a real development in the process, demonstrated by the awarding of the developer with an Environmental Award from the German Environmental Foundation in 2002.

Bischoff also deals with investigations of the Hydrobond system in [4.139]. A practical application example was in the VW foundry in Hannover where suction port cores were produced in conventional core shooting machines using binder content of between 1.3 to 1.7 %. The cores were placed automatically according to their production in the mould bales of a DISAMATIC form line, and the coring was carried out by a water jaket.

Figure 4.74: Wax pattern cluster for investment casting mould production (Image Dörrenberg stainless steel, Engelskirchen)

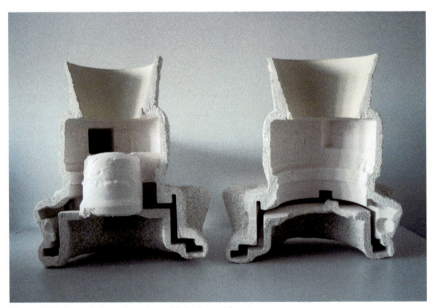

Figure 4.75: Ceramic investment mould shell mould, cut

The usage was deemed possible in principle because the attainable strength is higher than with PUR-cold box process and thus also making an automatic core loading feasible. Criticisms of the method were, long cycle times. *Hänsel* reports on the great resonance on the technical presentation of this method given at the first German meeting on the subject of inorganic binder systems that took place at the time [4.140]. Unfortunately, the further development of the method was less satisfactory than initially expected due to various unresolved problems in the course of the expansion of the Hydrobond binder system. Sometime later this process was supposedly used in the area of injection moulding of plastics. It is hoped that this or similar salt based binder systems may have a renaissance at some point as they certainly have potential!

LaempeKuhs® (LK) or BeachBox® process

The second inorganic mould material system based on inorganic binders that was developed from around the year 2001 is the LK (named after its two developers) also known as BeachBox® (so called on the basis of the use of tempered form raw materials). With this patented binder system two distinctions were made after two generations. In the so-called first generation of binders, the mixture was sulfate $MgSO_4 \times H_2O$ and mineral additives. According to *Bischoff* [4.141] a possible composition is 2–4wt% of anhydrous $MgSO_4$, 0.2–0.4wt% polyphosphate and 3–4wt% of water based on 100 % mould raw material.

The pure magnesium sulfate system, brought a number of problems, particularly with regard to humidity control and process safety under conditions of high humidity. This led to the parallel binder development and construction of an integrated binder-machine system called the BeachBox® process. With this machine technology the mould material preparation could be done in a pressure-tight mixer mounted over the core shooting machine whereby the form material could be heated to temperatures between 70 and 100 °C. By the pressure tightness of the mixer, premature solidification of the mixture by drying is prevented. The pre-heated mould material is shot into the hot core-production machine which is temperature controlled at 130–150 °C whereby the water in the core evaporates. This curing process can be accelerated by a purge of (cold or warm) air or through evacuation of the core box. The core boxes can be made of either aluminium or cast iron. To heat the tools, gas burners or electric heating elements are used.

In [4.142], microwave drying is specified as a possible variant solidification by *Gebhardt*, but in the aftermath the method was abandoned due to patent problems resulting in procedural similarities to the AWB® method.

Table 4.9 Examples of compositions of primary and secondary slurry [4.144]

Typical primary slurry for the investment casting mould production

	Zirconium silicate	Silica sol
Density [kg/l]	4,54	1,20
Mass [kg]	5,0	1,20
Volume [l]	1,10	1,0
Volume [%]	52,4	47,6
Bandwidth flour: 50 – 55 Vol.-% Binder: 45 – 50 Vol.-%		
Flour load: 4,5 – 5,5 kg Zirconium silicate, or 2,2 to 2,7 kg fused silica per 1 liter		
Additive: 0,2 – 0,5 Vol.-% wetting agent, 0,2 – 0,5 % defoamer (based on binder)		

Typical secondary slurry for investment casting mould production

	Chamotte	Silica sol	Water
Density [kg/l]	2,85	1,20	1,0
Mass [kg]	2,5	1,20	0,38
Volume [l]	0,88	1,0	0,38
Volume [%]	38,9	61,1	
Bandwidth flour: 37–44 Vol.-% Binder: 56–63 Vol.-%			
Flour load: 1,7–2,2 kg fireclay or 1,3–1,7 kg fused silica per 1 liter			

Due to the already discussed procedural difficulties, magnesium sulfate was replaced as the main binder component by second generation binders which were a mixture of mineral salts (sulfates, silicates, borates). These mineral constituents were distributed according to *Sobczyk* [4.136] to in the binder production under pressure as evenly as possible in a silicate-based binder. This basic binder has the general composition x $Me_2O * y SiO_2 * z H_2O$, which is very similar to a water glass solution. Thus, the original character of the binder was given up in the second salt binder system. At that time binder systems labeled LK 700 experienced considerable development.

The moulding mixtures prepared with the binders of this second generation reported good flow according to [4.142] so that the production of more difficult core geometries with thin walls was now possible. The better wettability of grains of the sand caused by nanoparticles in the binder is sadly not verifiable by respective sources. But due to this optimal wetting, a forming of a heat-resistant separating layer should be expected which prevents casting sand adhesion. The decoring of the castings produced in this way can be done with wet or dry methods. In the moulded part, gentle coring with water can be implemented and thereby structural improvements are achieved by the accelerated cooling. The waste sand mixture is then transferred to a reclamation unit. The binder should be reusable in full, but this is explicitly applicable only for the classic salt binder of the first generation. The silicate-based binder systems of the second generation are analogous to the classical water glass binder in that after reclamation there is no longer binding capability.

A dry decoring is possible according to [4.142] assuming that within the mould material system the binder will stay in the sand resulting in an increased binder effect. An example of the production of very complex cores through this method are the cylinder head cores produced at the Hannover VW foundry. Unfortunately, the development of the LK-method and the binder systems used for this purpose were discontinued some time later. It is hoped that, similar to the previously mentioned hydraulic bonding process, more developments will be pursued in the future. The pallet of available inorganic binder systems or moulding process could thus be extended. Mention is made of the method in the reports at the GIFA 2003 [4.143].

4.3.4 Investment casting with silica sol binders

The lost wax or investment casting method incorporates "lost" material, or sometimes wax for the pattern materials. These patterns are first assembled into so-called pattern clusters and mounted on a sprue. The future casting mould is built up around this pattern as a ceramic mould through alternate dipping in fireproof slurry (a mixture of binder and fireproof material) with subsequent sand-coating with coarser fireproof material and the drying of each coat. With sufficient shell thickness (in many cases 8 to 10 layers) and produced under the required conditions, this ceramic mould has sufficient strength to allow removal of the wax patterns through melting. The undivided ceramic mould gains its ultimate strength, and the properties required to absorb the liquid, in a firing process at temperatures of about 900 to 1000 °C. Subsequently, the still hot mould is poured off quickly. This is necessary because the manufactured precision castings usually have minimal wall thicknesses, and often difficult castable alloys – such as high alloy steels or nickel – are poured. While formerly there was only the alcoholic binder ethyl silicate available for the shell mould manufacturing, modern procedures today worldwide work almost exclusively with aqueous silica sol binders (silica). In both cases, the bond is formed through meshing of SiO_2 particles which adhere together by gelling or condensation in solutions, and in the process include other fireproof materials.

The following will be discussed in more detail regarding the individual steps of mould manufacturing technology in ceramic forms.

Slurry production

The mixture of silica sol binder and fire-proof material is called slurry. In addition to these two main components, it also contains other ingredients such as dispersants and de-foamers. As refractory raw materials, the otherwise rarely used silica powder is added in addition to finely ground zircon, mullite, chamotte, or silica sands. The average grain size comprising these flours usually range from about 2 microns to 60 microns. Sometimes nylon fibers are added into the slurry which burn in the subsequent firing of the ceramic mould. Through this, gas permeability is increased due to the resulting cavities in the shell mould. Considerable density differences exist in the slurry between the binder and the zirconium silicate (for example). The refractory raw materials, despite the addition of dispersants, tend to settle, and therefore the slurry container must be kept in constant motion. Because investment castings must have very low surface roughness, in the 1st to 3rd layer fine flour in the slurry is used and also sanding materials with small grain sizes surfactants are often used In the primary slurry. This allows for the complete wetting of the pattern clusters through the reduction of surface tension.

Immersion of the pattern clusters

The pattern clusters are then briefly dipped into a slurry bath. Here it is important that all surface parts are uniformly covered with the slurry material. In the next step it is important to ensure a steady drip off of excess slurry through targeted and reproducible moments of the pattern cluster outside of the slurry bath. For this reason, robots are often used in investment foundries.

Sanding

The slurry film applied to the patterns must be done evenly in the next step with refractory material. Here, the sprinkled material should have a very uniform particle size, since excessively fine pieces would lead to the development of uneven layers. Furthermore, the average particle size of the sanding material should increase inner to outwardly to achieve proper gas permeability (see table 4.10). The sanding material can be applied in two very different ways according to *Weihnacht* [4.146]. One method uses the "quasi-liquid" state of a fluidized sand bed in the phase when it is aerated from the bottom (eddy current bed or fluidizer). The other method is the uniform sprinkling of sand called "Rainfall". Both methods have technological advantages and disadvantages. In the fluidizer there is a danger of absorption of the fine particles. This is due to the fact that, because of the vertical flow of air, they settle on the sand surface and therefore can be absorbed by the wetted pattern cluster. Because of this, there is a greater possibility of the sand entering the cavities due to the fluid state. If the pattern cluster stays too long in the fluid bed there is a resulting phenomenon of "friction" of

Table 4.10: Example of sanding for an investment casting shell with 2 primary coatings

No. layer	1*	2*	3	4	5	6	7	8
AGS mm	0,17	0,17	0,18–0,5	0,18–0,5	0,18–0,5	0,5–1,0	0,5–1,0	0,5–1,0
Drying								

* - Primary coating

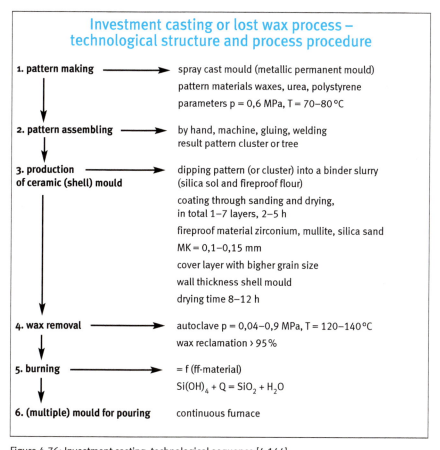

Figure 4.76: Investment casting, technological sequence [4.144]

the previously coated sand. Statistical assessments show that lower strengths result from using a fluidizer method. The "Rain-Sander" is applied for the even sprinkling of uniform sand grains. This requires that the sprinkled pattern cluster must be rotated and swiveled very carefully in order for the sand to reach all surface points, especially in the case of geometrically complex shapes.

The next step is for the applied slurry-refractory coatings to be dried. In the normal investment casting process, drying must be uniform and slow in order to prevent tension and the possibility of cracking in the shell moulds; the appropriate time frames can be seen in the representation of the process flow in figure 4.76. Therefore, an indoor environment with a temperature of between 22°C and 23°C and a relative humidity of 40 % is recommended. Innovative developments led to a new philosophy in drying ("cyclone technology") [4.147]. In this method, and depending on the layer coating, drying is carried out at a heat radiated (red light) temperature of 35°C to 40°C and 10 % humidity. While an accelerated air exchange is necessary with traditional technologies for generating a diffusion gradient within the slurry, the new technology uses a drying speed of only 8 m/s. The green strengths achievable due to the drying of the mould shells are also an influence on whether or not surface defects or cracks form during subsequent processes (wax melting, burning) [4.145, 4.148].

Wax melt
According to *Weihnacht* [4.146], an important aspect of the removal of the wax pattern or other wax components is the "thermal shock" method whereby the outermost layer

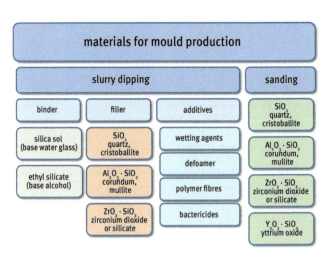

Figure 4.77: Investment casting slurries and sanding materials [4.145]

of wax (0.1 mm) is suddenly brought to a molten state in order to allow this wax to penetrate the porous ceramic and allow the next wax components to expand. This is reliably achieved through superheated high pressure steam treatment in the autoclave. This requires a completely dried ceramic mould so that the binding process between the colloids is completely irreversible, otherwise the moisturized water vapor would lead to dissolution of the binding. A rarely used wax removal method is the "Flash Fire" method which is done in a kiln heated to approximately 900 °C. Here, the wax is also abruptly brought to melt then allowed to flow through openings in the floor of the oven which makes it then necessary to employ an after burn furnace. In both of these wax melting methods it is necessary to ensure that the clusters are not heated above 25 °C before melting out. This is because there is a risk of wax expansion which can lead to the formation of hairline cracks and an "explosion" of the ceramic shell. Interesting aspects of wax melting with an autoclave are shown by Kügelgen in [4.149].

Burning of the shells
The final step for the production of ceramic moulds is the firing process. The rest of the residual wax must be removed (by burning) along with any moisture (by dehydration). The

Figure 4.78: Cyclone system for accelerated production of investment casting shell moulds [4.147]

Figure 4.79: Infrared drying of a coated pattern cluster [4.147]

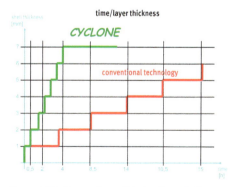

Figure 4.80: Comparison of drying times in the conventional shell manufacturing process and Cyclone method [4.147]

Warm or hot curing processes | 141

Figure 4.81: Shell structure of a investment casting shell mould after firing

mould is then pre-heated to ensure proper pouring temperature. The thermal time required for complete drying is often underestimated according to *Weihnacht* [4.146]. The last remaining moisture is eliminated at about 800 °C. Due to the low thermal conductivity of the ceramic mould and the direct influence of radiant heat on hard to reach areas of the ceramic mould, this temperature takes some time to be uniformly applied. This behaviour requires more attention when using aqueous binder systems compared to alcohol binders. A dwell time of at least one hour is recommended in a high temperature zone of around 950 °C. As a rule, pouring into the finished moulds takes place immediately upon removal from the kiln while it is still warm, but can also be done after cooling and renewed heating.

Literature-Chapter 4.3

[4.88] Svensson, I. L., Chemistry and Mechanical Properties of Carbon Dioxide cured Sodium Silicate Binders, Vortrag Nr. 4, 52, Internationaler Gießerei-Kongreß, Melbourne 1985
[4.89] Osterberg, L., Anderson, W., Stoberiet 62, 1985, Nr. 6, S. 8–16 und GIESSEREI 73, 1986, Nr. 22, S. 657
[4.90] Flemming, E., Schmidt, M., Hähnel, U., Ledwoin, E., Erfahrungen beim Einsatz von modifizierten Wasserglaslösungen, Gießereitechnik 35, 1989, Nr. 10, S. 299–302
[4.91] Döpp, R., Schneider, H., Jürgens, H., Eigenfeld, K., Keidis, A., Gießversuche zum Wasserglas-CO_2-Verfahren – Teil 2, Betriebsversuche, GIESSEREI 76, 1989, Nr. 23, S. 799–802
[4.92] Döpp, R., Schneider, H., Fortschritte des umweltfreundlichen Wasserglas-CO_2-Verfahrens zur Kern- und Formherstellung, Gießerei-Erfahrungsaustausch 1990, Nr. 11, S. 465–467
[4.93] Doroschenko, S. P., Makarevitsch, A. P., Formsande und Kernsande mit Wasserglasbindern, Probleme und Perspektiven, Vortrag 61. Gießerei-Weltkongreß, Peking 1995
[4.94] Marachovec, L. N., Lemkow, J., Heißhärtende Formstoffmischungen mit modifiziertem Wasserglasbindemittel, Litejnoe Proizvodstvo, 1997, Nr. 10, S.12
[4.95] Müller, J., Informationen zum Bindersystem Inotec und Angabe wichtiger Literaturstellen, persönliche Mitteilung 2011

[4.96] Müller, J., Weicker, G., Körschgen, J., Serieneinsatz des anorganischen Bindemittelsystems INOTEC im Leichtmetallguss, GIESSEREI-PRAXIS 05/2007, 192–194
[4.97] Müller, J., Koch, D., Frohn, M., Weicker, G., Körschgen, J., Schreckenberg, S., Inotec bewährt sich in der Praxis, GIESSEREI 95, 2008, Nr. 1, S. 44–48
[4.98] Wallenhorst, C., Grundlagen zum Verständnis der anorganischen Kernfertigung, GIESSEREI-PRAXIS 2010, Nr. 6, S. 181–184
[4.99] Wallenhorst, C., Körschgen, J., Frohn, M., Kasperowski, A., Steuerung der Prozessstabilität bei der anorganischen Kernfertigung, GIESSEREI 97, 2010, Nr. 12, S. 58–61
[4.100] Weissenbeck, E., Willimayer, J., Wolf, J., BMW-Leichtmetallgießerei setzt auf anorganisch gebundene Kerne, GIESSEREI 95, 2008, Nr. 6, S. 32–35
[4.101] Pabel, T., Kneißl, C., Brotzi, J., Müller, J., Anorganisches Bindersystem: Einsatz von INOTEC-Kernen für deutlich verbesserte mechanische Eigenschaften von Al-Gussteilen, GIESSEREI-PRAXIS 2009, Nr. 11, S. 359–366
[4.102] Kautz, T., Weissenbeck, E., Blümlhuber, W., Anorganische Sandkernfertigung: ein Verfahren mit Geschichte, GIESSEREI 97, 2010, Nr. 8, S. 66–69
[4.103] Weissenbeck, E., Kautz, T., Brozki, J., Müller, J., Zylinderkopffertigung der Zukunft – Ökologie, Ökonomie und Werkstoffoptimierung im Einklang, MTZ 06/2011, 72. Jahrgang, 484–489
[4.104] Schwickal, H., Hoffmann, H., Blümlhuber, W., Weissenbeck, E., Regenerierung von anorganisch gebundenen Gießereikernsanden, GIESSEREI 96, 2009, Nr. 11, S. 40–44
[4.105] Sasse, S., Knechten, J., Brotzi, J., Wallenhorst, C., Wojtas, H.-J., Entwicklung eines anorganischen Bindersystems für die GJL-Bremsscheibenfertigung GIESSEREI 98, 2011, Nr. 4, S. 36–40
[4.106] Birnbaum, C., Entwicklungen bei anorganischen Formstoffsystemen im Zeitraum 1998–2008, Literaturarbeit, TU BA Freiberg, 2009
[4.107] Gießerei Hannover: Anorganisches Mittel für den Serienguss, GIESSEREI 91 (2004), Nr. 8, S. 8
[4.108] Voigt, P., Bischoff, U., Ristau, B., Georgi, B., Still, H., und Lustig, C., Einsatz neuer anorganischer Formstoffbindemittel in der Produktion von Aluminium-Zylinderköpfen, GIESSEREI 93 (2006) Nr. 3, S. 44–49
[4.109] Löchte, K., und Boehm, R., Cordis, Das anorganische Bindemittelsystem – Eigenschaften und Erfahrungen, GIESSEREI 92 (2005) Nr. 3, S. 68–72
[4.110] Bischoff, U., und Voigt, P., Mit neuen anorganischen Formstoffbindern auf dem Weg zur Serienreife im Aluminiumguss, Giesserei-Rundschau (2006) Nr. 9, S. 191–193
[4.111] Grassmann, H.-C., Hochfrequenztrocknung wasserglasgebundener Kerne, GIESSEREI 49, 1962, Nr. 17, S. 580–582
[4.112] Schreyer, G. W., Erbs, H.-D., Hoefer, H., Herstellung von Gießereikernen, VEB Deutscher Verlag für Grundstoffindustrie, Leipzig, 1965
[4.113] Tripsa, I., Dragoi, F., Cosneanu, C., Tarasescu, M., Kerntrocknung durch Mikrowellen, 43. Internationaler Gießerei-Kongreß, Bukarest, 1976, Vortrag Nr. 33
[4.114] Cole, G. S., Nowicki, R. M., Owusu, Y. A., Microwave Cured Sodium Bonded Cores, AFS Transactions, 1979, S. 605–612
[4.115] Owusu, Y. A., Draper, A. B., Inorganic Additives Improve the Humidity Resistance and Shakeout Properties of Sodium Silicate Bonded Sand, AFS Transactions 1981, S. 47–54
[4.116] Owusu, Y. A., Sodium-Silicate Bonding in Foundry Sands, Dissertation, Pennsylvania State University 1980 und GIESSEREI 72, 1985, Nr. 4, S. 93
[4.117] Owusu, Y. A., Draper, A. B., Nowicki, R. M., Cole, G. S., Humidity Resistance of Sodium Silicate Bonded Sands Cured with Microwave Energy; AFS Transactions 1980, S. 601–608

[4.118] Ailin-Pyzek, I. B., Spencer, R. W., Falcone, J. S., Silicate Foundry Binders with Improved Humidity Resistance, AFS Transactions 1981, S. 543–546
[4.119] Carlson, G., Thyberg, B., Core production by the sodium silicate warm-air process, Vortrag Nr. 23, BCIRA International Conference, Coventry/Birmingham, 1986
[4.120] Chang, H., Chen, E. L., Lindeke, R., AFS Transactions 96, 1988, S. 217–222 und GIESSEREI 78, 1991, Nr. 22, S. 827
[4.121] Flemming, E., Mende, H., Beitrag zu Problemen der Qualität und Verwendung von Wasserglaslösungen als Formstoffbinder, Gießereitechnik 35, 1989, Nr. 6, S. 192–197
[4.122] Verfahren zur Herstellung einer Gießform; Europäisches Patent Nr. WO 89/02325
[4.123] Jelinek, P., Contribution of the Czechoslovak Foundry Industry to Chemization of Manufacture of Molds and Cores on the Base of Alcali Silicates, Slevarenstvi, 1996, Nr. 2, S. 85–103
[4.124] Heat-Cured Foundry Binders and Their Use, Weltpatent WO 95/11787, 1995
[4.125] Xiao, B., Xu, Z., Xiang, X., The Microwave Waterglass Process and the Reclamation of Waterglass Bonded Sands, CIATF-Kom. 1.6, Vortrag am 24.09.1994, Peking
[4.126] Döpp, R., Alekassir, A., Brümmer, G., Erweiterung des arbeitsplatz- und umweltfreundlichen Wasserglas-Formverfahrens zur Form- und Kernherstellung in Gießereien, Abschlussbericht Forschungsvorhaben 01 VQ 934 A/1, TU Clausthal, 1995
[4.127] Alekassir, A., Beitrag zum Mikrowellen-Trocknen und zum Regenerieren wasserglasgebundener Formstoffe, Dissertation, TU Clausthal, 1997
[4.128] Piegiel, M., Granat, K., Zastosowanie Mikrofalowego Nagrzewania W Odlewnistwie, Solidification of Metals ans Alloys, 1997, Nr. 33, Katowice, S. 248–254
[4.129] Piegiel, M., Microwave Hardening of Waterglas Sandmixes, Vortrag Konferenz anlässlich des 50jährigen Jubiläums des Einsatzes von Wasserglas in der Gießerei, Zlin, 20.–21.10.1998
[4.130] Polzin, H., Untersuchungen zur Mikrowellenverfestigung von wasserglasgebundenen Gießereiformstoffen, Freiberger Forschungsheft Nr. B 302, 2000
[4.131] Polzin, H., Flemming, E., Untersuchungen zur Mikrowellenverfestigung von wasserglasgebundenen Gießereiformstoffen, Teil 1: Versuchsergebnisse zu erreichbaren technologischen Eigenschaften und praktische Erprobung, GIESSEREI-PRAXIS 1999, Nr. 12, S. 569–581
[4.132] Polzin, H., Flemming, E., Untersuchungen zur Mikrowellenverfestigung von wasserglasgebundenen Gießereiformstoffen, Teil 2: Ansätze zur Strukturaufklärung der bei der Mikrowellentrocknung ablaufenden Vorgänge, GIESSEREI-PRAXIS 2000, Nr. 2, S. 58–71
[4.133] Wolff, A., Steinhäuser, T., AWB – ein umweltverträgliches Kernherstellungsverfahren, GIESSEREI 91 (2004), Nr. 6, S. 80–83
[4.134] Gosch, R., Die Entwicklung eines Kernfertigungssystems auf anorganischer Binderbasis zur Serienreife – Innovation und Nachhaltigkeit in idealer Umsetzung, Giesserei-Rundschau, 2004, Nr. 51, S. 139–142
[4.135] Steinhäuser, T., Wehren, B., Wiesauer, J., Einsatz anorganischer Binder in der neuen Kokillengießerei der MWS Castings s.r.o., Slovakei, GIESSEREI 95 (2008), Nr. 6, S. 66–68
[4.136] Sobczyk, M., Untersuchung zur Nutzung der Vakuumtrocknungshärtung für die Herstellung und den Einsatz magnesiumsulfatgebundener Kerne für den Leichtmetallguss, Dissertation Universität Magdeburg, 2008
[4.137] Tilch, W., GIFA 2003: Form- und Kernherstellung (Teil 1), GIESSEREI 93 (2003), Nr. 10, S. 42–44
[4.138] Hänsel, H., Ein neues Bindersystem der innovativen Art, Teil 1, Das Verfahren und dessen Einsatz in der VW-Gießerei Hannover unter Serienbedingungen, GIESSEREI 89 (2002), Nr. 2, S. 74–76

[4.139] Bischoff, U., Untersuchungen zum Einsatz anorganischer, wasserlöslicher Kernbindersysteme in einer Aluminium-Leichtmetallgießerei, GIESSEREI-PRAXIS (2002), Nr. 12, S. 454–455
[4.140] Hänsel, H., Ein anorganisches Bindersystem der innovativen Art, VDG-Fachtagung „Anorganische Binder – Durchbruch oder ewige Hoffnung?", Wuppertal, 14.11.2002
[4.141] Bischoff, U., Untersuchungen zum Einsatz anorganischer, wasserlöslicher Kernbindersysteme in einer Aluminium-Leichtmetallgießerei, Dissertation, TU Bergakademie Freiberg, 2003
[4.142] Gebhardt, S., Kuhs, B., Rautenbach steht kurz vor der Serieneinführung eines Zylinderkopfes mit LaempeKuhs-Binder, GIESSEREI 92 (2005), Nr. 5, S. 68–69
[4.143] Tilch, W., Polzin, H., GIFA 2003 – Formstoffe, Formverfahren und Maschinen zur Form- und Kernherstellung, Formstoffaufbereitung und Regenerierung, GIESSEREI-PRAXIS 2003, Nr. 10, S. 409–410
[4.144] Polzin, H., Vorlesungsscript Formverfahren II, Technische Universität Bergakademie Freiberg, 2011
[4.145] Rothe, H., Untersuchung der Rissbildung bei der Herstellung keramischer Feingußschalenformen, Dissertation, RWTH Aachen, 1999
[4.146] Weihnacht, W., Dokumentation zur Herstellung von keramischen Schalenformen für das Feingießverfahren, 2008
[4.147] Weihnacht, W., Eine neue Philosophie des Feingießens, Vortrag 16. Ledebur-Kolloquium TU Bergakademie Freiberg, Gießerei-Institut, 26. und 27. Oktober 2006
[4.148] Meulenberg, W., Werkstoffverhalten und Rissentstehungsursachen von Feingussformschalen aus Zirkon-Mullit-SiO_2-Kombinationen, Dissertation, RWTH Aachen, 1999
[4.149] Kügelgen, M., Autoklav Next Generation (NG) – Neue Wege für das schonende und schnelle Wachsausschmelzen von keramischen Formschalen im Feinguss, GIESSEREI-PRAXIS 2011, Nr. 3, S. 94–99

5 The Use of Alternative Moulding Materials

For the production of lost moulds and cores one requires moulding materials which consist of the main components of base moulding material, moulding material binder and hardener, and sometimes various moulding material additives and auxiliary materials. The main component of each moulding material is the base moulding material, which is approximately between 85 % and 98 %. The base moulding material (the 'sand') is the main component of the moulds system. For the production of lost moulding moulds and cores, natural or synthetic sand grains can be used. Minerals in diameter of between 0.02 mm and 2 mm are recommended. Moulding material raw materials are made up of mixed solid grains with specific granulometric properties and form the skeleton of the moulding material or core. There are generally three property complexes for foundry technology as shown in table 5.1.

The most important moulding material used today is silica sand, a crystalline form of silica with the main component of SiO_2. This, according to *Blankenburg* [5.1] and others was created in the course of earth history through weathering of older rock containing quartz, and is obtained in open-pit mines using either wet or dry mining techniques. Silica sand is available in large qualities and in good quality in central Europe which forms the basis for wide use in the foundry industry. However, silica sand also has some negative characteristics which can lead to problems in the production of high quality castings. Firstly, there is the expansion behaviour. Silica sand makes three modification changes with specific temperatures rises (573 °C, approx. 870 °C and 1400 °C) including density and length change. This transition happens in virtually every casting operation, while two of the transitions (irreversible) are dependent on the material to be cast. Such conversions can cause expansion errors (scabbing, veins, rat tails) that disintegrate the moulding material due to tensions which exceed the mould material strengths, and subsequently, liquid metal penetrates into the emerging cracks or cavities. Another drawback of the silica sand is its high temperature characteristic which is expressed by the initial sintering temperature. High quality silica sand begins to sinter at 1450 °C to 1550 °C (superficial melting of the sand grains). With larger amounts of impurities (such as alkali oxides) sintering may start at significantly lower temperatures. This leads to problems in the production of high melting alloys like steel cast iron, which can manifest in the form of defects such as burning, sintering or mineralization. A third drawback of the silica sand which should be mentioned is its thermo-physical properties, such as poor thermal conductivity that can lead to difficulties in the production of thick-walled cast iron or steel parts.

The disadvantages mentioned here lead to the use of alternative mould basic substances that often do not contain silica moulding material base materials. There are now a number of such alternative forms of raw materials for use in moulding material and core production. In addition to the classic sands used such as chromite, zircon, olivine sand or chamotte, there are now a number of synthetically generated materials with largely definable grain size and shape, for example, alumino silicate sands (mullite) and corundum.

Table 5.1: Property complexes for technological evaluation of foundry moulding materials [3.7]

Property complex	Individual features
Chemical and mineralogical composition	Chemical composition, Proportion of base component, for example, SiO_2 content, further components
	Real structure of the grains
	Chemistry of the grain substance or the grain surface
Granulometric and morphological parameters	Particle size and particle size distribution average grain size, fine grain
	Grain shape
	Structure of the surface of representative grains
	Specific surface area of the grain bulk material
	Chemical 'activity' of the grain surface
Physical and technological properties	Hardness of representative grains
	Density of representative grains
	Expansion behavior of the grain bulk materials
	Sintering behaviour of grain bulk materials
	Crushing behaviour of the bulk material
	(Slope of the grains to burst at mechanical or thermal stress)
	Thermal properties such as thermal conductivity

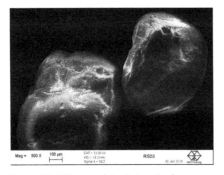

Figure 5.1: Silica sand, typical grain shape and structure

Figure 5.2: Silica sand with binder shells

The use of alternative forms of moulding material with inorganic binders

According to *Beckius* in [4.68], one of the oldest applications of alternative forms of raw materials for inorganic moulding material binder in the foundry is the use of olivine sand for moulding material and core production with water glass binder systems. Olivine sand is found mainly in Scandinavia and is made up of magnesium silicates which have strong basic characteristics and therefore, do not harden to form acidic mould material binders such as furan resins. The deposits of these sands in Northern Europe provide an explanation for the widespread practice of water glass ester moulding material in Scandinavia today. *Beckius* goes on to say in [4.68] that the use of water glass bonded olivine mould materials compares with the resin process (which uses other moulding material base materials) in achievable strength properties. If one con-

Figure 5.3: Chromite sand with binder shells

Figure 5.4: Synthetically produced aluminosilicate (mullite) with binder shells

siders olivine and silica sand together with hardening by water glass, higher primary strength and an improved decay behaviour can be observed. *Jelinek* reports on the long tradition of the use of chromite sands with water glass binder in Czech foundries [4.123]. Thus, the moulding material made of crushed chromium ore is particularly suitable to produce large and heavy steel castings requiring high quality. Through the use of chromite sands, defects such as expansion defects or burns can be reduced or avoided. Furthermore, water glass ester bonded chromite sand moulds have better disintegration behaviour than those made with silica sand. The better cooling properties of chromite sand are also discussed in *Jelinek's* article. In this context, it must be pointed out explicitly that the effect of this raw moulding material with regard to firing or penetration is related to the better heat conductivity and not with sintering behaviour. This means that chromite sand should only be used to improve unpacking behaviour and casting surfaces in the instances where heat can be rapidly removed from the casting. If, for example, the cores are only connected to the form by small core markers, it is possible that due to the insufficient heat dissipation during the heating of the core material and the low sintering start temperature, considerable scorching effects can take place, even as far as mineralization. *Nürnberg* does a comprehensive work on this subject in [4.48]. He examines thirteen different commercially available alternative raw moulding materials pertaining to the water glass CO_2 process, the water glass ester process and the water glass hot box process. Table 5.2 presents the materials used. Tables 5.3 and 5.4 give the granulometric structure of the sands and some important property values. The studies are divided into three main areas: The determination of the primary strengths, the decay behaviour (by measuring residual strengths) and the warm formability which is measured by hot distortion testing.

Water glass CO_2 process

The starting point reference presented in all described studies for assessing the investigated alternative basic material is silica sand H32. In the studies, the point discussed earlier here was made clear that with increased gassing times immediate strengths usually rise within a core storage time of 24 hours. After that less strength falls off than with lower gassing times. Figure 5.8 shows that with a binder addition of 3 % and a gassing period of 10 seconds (here the lowest gassing time), bending strengths are reproduced. It is based on silica sand with a 24 hour strength of about 140 N/cm^2 whereby the immediate strength is not measurable with low gassing times. Chromite, zircon sand and bauxite all have comparable strengths to silica sand, while the other investigated moulding base materials are not appropriate due to the low strength. If one considers the decay behaviour shown in figure 5.9, one finds that the three sands acceptable in terms of the primary strengths have primarily an unfavorable decay behaviour. However, it must be noted that the process for determining the residual strength used here may be only appropriate for a clear assessment under certain conditions. The bending strength test specimens were stored for 24 hours and then submitted to a test temperature of 400 °C or 800 °C for 2 minutes. They were then tested after cooling to room temperature. This test process can also be applied when comparing decay behaviour in the application of different raw mould materials. It can also be applied in the three different curing technologies discussed here.

Water glass ester process

The strength behaviour of the investigated moulding base materials with self-hardening water glass binders is shown in figure 5.10. The binder contents in these experiments were 3 % and the hardener levels were at 0.3 %. The reference samples with silica sand H32 stayed in a conventional framework between 200 N/cm² and 250 N/cm². In contrast to the gassing with carbon dioxide there were no significant

Table 5.2: Chemical and mineralogical composition of the studied raw moulding materials [4.48], [5.2], [5.3]

Moulding material	Components	Mineralogical composition
H 32	Silica sand	› 99 % SiO_2
M-sand	Aluminosilicate ceramic	› 76 % Al_2O_3 ; 23,5 % SiO_2
J-28	Silica-feldspathic sand	78-82 % Al_2O_3 ; 8-12 % SiO_2
R-sand	Rutile sand	96 % TiO_2 ; 0,9 % Fe_2O_3
Cerabeads	Aluminosilicate sand	60-62 % Al_2O_3 ; 36-38 % SiO_2
Natural olivine	Mineral sand	47,5 % MgO ; 41,6 % SiO_2 ; 8 % Fe_2O_3
Fireclay (chamotte)	Fireclay refractory bricks	42 % Al_2O_3 ; 54 % SiO_2
SiC	Ceramic oxide	99 % SiC
Chromite sand	Chromite ore	46 % Cr_2O_3 ; 27,5 % FeO ; 15 % Al_2O_3 10 % MgO ; 1,5 % SiO_2 ; 0,5 % TiO_2
Kerphalite	Min. aluminosilicate sand	60-62 % Al_2O_3 ; 38-40 % SiO_2
Reg. alumina	Oxide ceramic	95 % Al_2O_3
Zircon sand	Heavy mineral	65 % ZrO_2 + HfO_2 ; 32,5 % SiO_2
Bauxite sand	Aluminosilicate melt	80-82 % Al_2O_3 ; 8-12 % SiO_2 ; 3–5 % Fe_2O_3 ; 3–5 % TiO_2

anomalies here with the exception of the fireclay. There were also a number of sands that behaved similarly to silica sand. The example of olivine sand demonstrates why this raw moulding material, which can only be processed successfully with alkaline binder systems, is used successfully in the Scandinavian countries along with other comparable binder contents. Corundum, zircon sands and bauxite show significantly higher strengths with the water glass ester process. Correspondingly higher residual strengths are to be expected in this process variant (figure 5.11). Regarding temperature stress there is no uniform picture here, i.e. some of the sands used are more likely at lower (pouring) temperatures to have an improved decay while others behave similarly at higher temperature loads.

Water glass hot box process

Images 5.12 and 5.13 show the primary and secondary resistance behaviour of alternative raw moulding materials in hot box curing. When considering strength it is clear

Table 5.3: Granulometric and morphological parameters of the investigated alternative sands [4.48], [5.3]

Moulding material	Grain shape	AFS No.	Grain size fractions in mm and %						
			>0,5	>0,355	>0,25	>0,18	>0,125	>0,09	>0,063
H 32	Edge rd.	44	2	28	56	12	2	0	0
M-Sand	Angular	60	0	13	26	32	19	8	2
J-28	Edge rd.	55	3	21	33	25	13	4	1
R-Sand	Rounded	75	0	0	3	37	52	6	2
Cerabeads	Round	65	0	0	24	46	25	5	0
Olivine sand	Angular	45	4	16	26	23	18	9	4
Fireclay (chamotte)	Angular	47	0	37	36	18	9	0	0
SiC	Angular	46	1	35	24	24	16	0	0
Chromite sand	Edge rd.	58	5	17	32	25	14	5	2
Kerphalite	Angular	60	0	4	30	36	26	4	0
Reg. alumina	Edge rd.	40	1	62	37	1	0	0	0
Zirkon sand	Rounded	70	0	0	3	37	52	6	2
Bauxite sand	Rounded	60	0	4	30	36	26	4	0

Table 5.4: Physical and technological parameters of the investigated alternative moulding materials [4.48], [5.3], [5.4]

Moulding material	Pure density g/cm³	Bulk density g/cm³	$l_A k$** $10^{-1} K^{-1}$ 20-800 °C	WAV *** %	T_{Sinter} °C
H 32	2,65	1,4	17,0	0,77	›1550
M-sand	3,1	1,5	4,8	0,25	1250
J-28	2,6	1,51	11,5	0,45	1150
R-sand	4,25	2,38	8,3	0,62	1400
Cerabeads	2,9	1,59	4,3	1,11	1250
Olivine sand	3,3	1,61	11,0*	–	1450
Fireclay (chamotte)	2,5	1,23	4,8*	1,46	1250
SiC	3,21	1,62	4,7*	0,19	–
Chromite sand	4,5	2,73	7,5	0,15	750
Kerphalite	3,1	1,53	7,0	0,84	1250
Regular alumina	3,98	1,87	7,2*	0,31	900
Zirkon sand	4,5	2,81	4,5	0,18	1350
Bauxite sand	3,3	2,08	7,4	0,27	1100

* Expansion coefficient 20-600 °C
** Coefficient of linear expansion
*** Water suction capacity

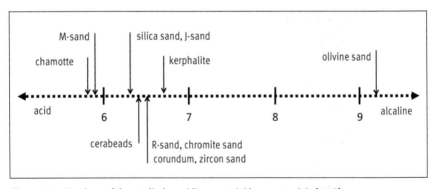

Figure 5.5: pH values of the studied moulding material base materials [4.48]

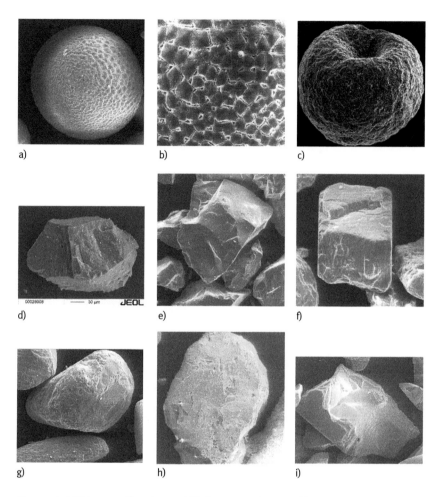

Figure 5.6: SEM images of bauxite sand (a), bauxite sand surface (b), Cerabeads (c), M-sand (d) Kerphalite KF (e), J-sand (f), R-sand (g), fireclay (h) and cubic SiC (i) [5.5]

that the hardening temperature in this area has only a minor role. This is also evident for the hot curing moulding material systems discussed in previous sections that work with core box temperatures of 140 °C to 160 °C. Therefore, the decay behaviour in figure 5.12 is only shown for a temperature of 150 °C. Chamotte has significantly poorer performances when compared to silica sand. Also interesting is that although olivine sand is very suitable for the water glass ester process, it shows a rather nega-

Alternative Moulding Materials | 153

Figure 5.7: grain collective [5.5] of –

J-sand

Kerphalite KF

Cerabeads

M-sand

Bauxite sand

R-sand

Fireclay

SiC

154 | Alternative Moulding Materials

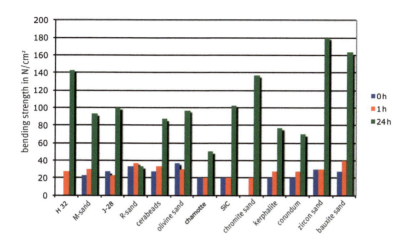

Figure 5.8: Water glass CO_2 process, bending strength, 3 % binder, 10 s gassing [4.48]

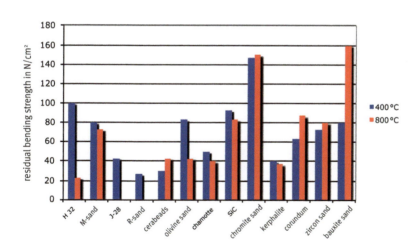

Figure 5.9: Water glass CO_2 process, residual strengths, 3 % binder, 10 s gassing [4.48]

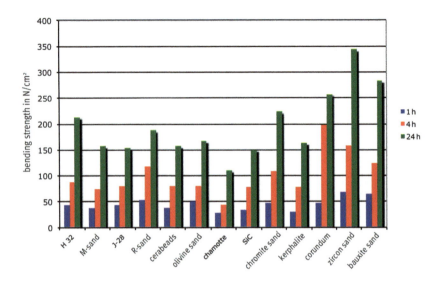

Figure 5.10: Water glass ester process, bending strength, 3 % binder, 0.3 % hardener [4.48]

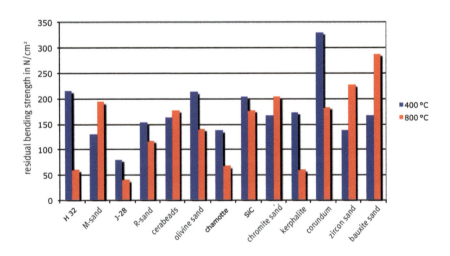

Figure 5.11: Water glass ester process, residual strengths, 3 % binder, 0.3 % hardener [4.48]

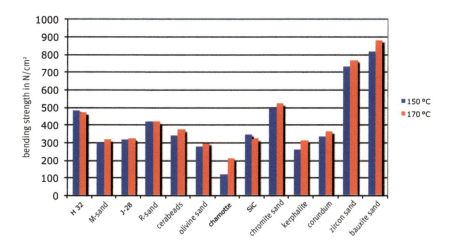

Figure 5.12: Water glass hot box process, bending strength, 3 % binder, 1 min baking time, test after 1 h storage time [4.48]

tive behaviour in thermal curing. Chromite, zircon and bauxite sands reach significantly higher bending strengths in hot box curing than the silica sand moulding mixtures. The residual strengths adapt to the primary strengths, as evidenced in figure 5.13. However, it can be stated here that the residual strengths are generally lower at 800 °C and (significantly) than at 400°C. The reason for this may be the fact that in this variant, a purely physical curing is achieved by drying. Due to the fact that with the hot box process, chemical reactions do not run parallel to other curing variants so one can, in principle, calculate decay behaviour improvement. By the absence of reaction by-products such as sodium carbonate or acetate, the development of a second maximum strength through glass phase formation is suppressed and thus provides the illustrated positive results.

Hot distortion (hot formability)

The hot formability or hot distortion test is originally from the quality control of shell moulding materials. By this test, process results can be examined regarding time dependent deformation, and ultimately, the fracture behaviour of different moulding materials. To evaluate the thermal plasticity of the cured moulding material, a special cantilevered specimen is heated from the bottom at a temperature of 1000 °C. From the

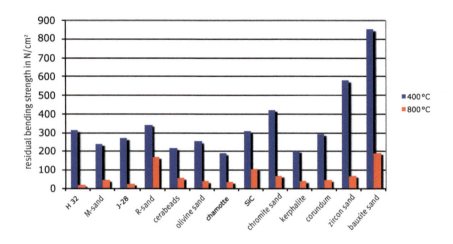

Figure 5.13: Water glass hot box process, residual strengths, 3 % binder, baking time 1 min at 150 ° C [4.48]

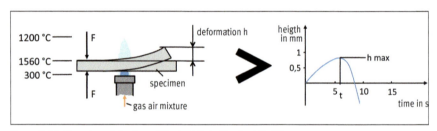

Figure 5.14: Experimental setup of the hot distortion test [4.48]

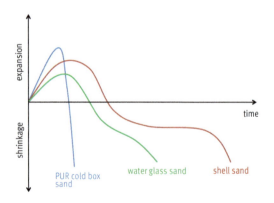

Figure 5.15: Hot deformation curves of different moulding materials

158 | Alternative Moulding Materials

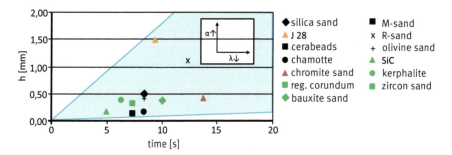

Figure 5.16: Hot distortion test on different moulding materials [4.48]

obtained time-distance curve, conclusions can be drawn on whether the inspected material is a thermoplastic moulding material (e.g. shell moulding material) or a very brittle (e.g. pure cold box) fracture behaviour. To avoid certain casting defects such as veining, thermoplastic behaviour is desirable. Figure 5.14 shows the schematic sequence of hot-distortion testing and in figure 5.15 the differences between brittle and plastic moulding materials are clear.

As for the strength tests, the tested raw moulding materials show different results with respect to the hot deformation behaviour (figure 5.16): In all specimens, the characteristic curve for water glass moulding materials was evident. The amount of deformation is related to the expansion coefficient. Olivine, J- and R-sands showed higher or comparable displacement to silica sand. Smaller displacements are therefore, achieved by moulding materials with a poorer expansion behaviour compared to silica sand. Mainly sands of aluminium oxides (cerabeads, fireclay) or highly thermally conductive materials (zirconium oxide, SiC) have lower expansion coefficients. The thermal conductivity often behaves contrary to the linear expansion behaviour, i.e. substances with low heat conduction use the introduced energy for expansion and vice versa. For use in the foundry, moulding materials that have low displacement and longer plastic phases are desirable because this prevents expansion defects such as veining or crusting while maintaining dimensional stability. Of all the examined raw moulding materials, chromite and bauxite sand have significantly better values compared to silica sand.

Table 5.5: Summary of the properties of the investigated raw moulding materials [4:52]

Moulding material	AGS	S_{th}	Grain shape	pH	ρ_S	α	WAV	σ_B	$\sigma_R^{400°C}$	$\sigma_R^{800°C}$
Waterglass CO_2 process										
H 32	+	--	Edge rounded	-	--	++	+	++	++	--
M-sand	-	+	Angular	--	-	--	--	-	+	-
J-28	+	+	Edge rounded	-	-	+	-	+	--	--
R-sand	--	++	Rounded	+	+	-	-	--	--	--
Cerabeads	-	++	Round	-	-	--	++	-	--	-
Olivine sand	+	-	Angular	++	-	+	/	-	+	-
Fireclay (chamotte)	+	-	Angular	--	--	-	++	--	-	-
SiC	+	-	Angular	/	-	-	--	+	+	+
Chromite sand	+	+	Edge rounded	+	++	-	--	++	++	++
Kerphalite	-	+	Angular	+	-	-	+	-	--	--
Standard corundum	++	--	Round	+	-	-	-	-	-	+
Zirkone sand	-	++	Rounded	+	++	--	--	++	-	+
Bauxite sand	-	+	Rounded	/	+	-	--	++	+	++
Water glass ester process										
H 32	+	--	Edge rounded	-	--	++	+	+	+	--
M-sand	-	+	Angular	--	-	--	--	-	--	+
J-28	+	+	Edge rounded	-	-	+	-	-	--	--
R-sand	--	++	Rounded	+	+	-	-	-	-	-
Cerabeads	-	++	Round	-	-	--	++	-	-	+
Olivine sand	+	-	Angular	++	-	+	/	-	+	-
Fireclay (chamotte)	+	-	Angular	--	--	-	++	--	--	--
SiC	+	-	Angular	/	-	-	--	-	+	+
Chromite sand	+	+	Edge rounded	+	++	-	--	+	-	+
Kerphalite	-	+	Angular	+	-	-	+	-	-	--
Standard corundum	++	--	Round	+	-	-	--	+	++	+
Zirkone sand	-	++	Rounded	+	++	--	--	++	--	++
Bauxite sand	-	+	Rounded	/	+	-	--	++	-	++

Continuation Table 5.5

Moulding Material WGL-Warmbox-Verfahren	AGS	S_{th}	Grain shape	pH	ρ_S	α	WAV	$\sigma^{170°C}$	$\sigma_R^{400°C}$	$\sigma_R^{800°C}$
H 32	+	--	Edge rounded	–	--	++	+	+	+	--
M-Sand	–	+	Angular	--	–	--	--	–	–	--
J-28	+	+	Edge rounded	–	–	+	–	–	–	--
R-sand	--	++	Rounded	+	+	–	–	+	–	+
Cerabeads	–	++	Round	–	–	--	++	–	–	--
Fireclay (chamotte)	+	-	Angular	--	--	–	++	--	--	--
SiC	+	-	Angular	/	–	–	--	–	–	+
Chromite sand	+	+	Edge rounded	+	++	–	--	++	++	–
Kerphalite	–	+	Angular	+	–	–	+	–	--	--
Standard corundum	++	--	Round	+	–	–	--	–	+	--
Zirkone sand	–	++	Rounded	+	++	--	--	++	++	–
Bauxite sand	–	+	Rounded	/	+	–	--	++	++	++

Legend:
AGS – average grain size
S_{th} – specific surface
ρ_S – bulk density
α – coefficient of linear expansion
WAV – water suction capacity
σ_B – bending strength at room temperature
$\sigma^{170°C}$ – bending strength of hot box process at 170 °C core box temperature
$\sigma_R^{400°C}$ – residual bending strength at 400 °C
$\sigma_R^{800°C}$ – residual bending strength at 800 °C

++ very good
+ good
/ neutral
– poor
-- very poor

Summary

In summary, it can be stated that the fundamentally differing technological behaviour of the tested variants of water glass moulding materials can be naturally applied to the investigation of alternative raw materials. A clearly worse or better performance compared to the silica sand cannot be proved with many of the alternative sands that were considered. Chromite, corundum, zircon and bauxite sands tend to have better primary strength characteristics. This applies also to some extent to the thermal deformation behaviours. Raw moulding materials containing aluminium oxide generally have lower strengths than silica sand. Regarding the secondary strength or decay slope there are generally other sands which show a better performance than silica sand. That these raw moulding materials also have a lower level of output resistance is apparent from the considerations in this section. The results show that the inorganic moulding material binder systems are also competitive in terms of the use of alternative raw moulding materials. For specific application one must take into account the targeted property levels (e.g. strength or decay behaviour) as well as the prevention of substance-specific casting defects in order to gain an optimization of binder content and curing technology. Table 5.5 summarizes the varying results.

Literature-Chapter 5

[5.1] Blankenburg, H.-J., Quarzrohstoffe, Dt. Verlag für Grundstoffindustrie, 2. Auflage, 1992
[5.2] Tilch, W., Martin, M., Brinschwitz, R., Polzin, H., Einfluss alternativer Formgrundstoffe auf die Eigenschaften von Formstoff und Gussteil, GIESSEREI 2006, Nr. 8, S. 12–24
[5.3] Produktinformation HA-Spezialsande, Hüttenes Albertus GmbH, 2008
[5.4] Recknagel, U., Dahlmann, M., Spezialsande, GIESSEREI 2007, Nr. 6, S. 190–197
[5.5] Recknagel, U., Dahlmann, M., Spezialsande – Formgrundstoffe für die moderne Kern- und Formherstellung, GIESSEREI-PRAXIS, 2010, Nr. 11, S. 346–353

6 Reclamation of Used Sands

Basics of reclamation

Reclamation and re-use of old moulding materials in the production of moulds and cores are imperative for economic and ecological reasons. Reclamation of used sands is defined as 'The conversion of foundry used sands into a moulding material with properties similar to new sands by removal of spent residual binder.' Thus, the reclamation differs fundamentally from the reprocessing of bentonite-bonded moulding materials. While simply replacing the used elements (binder, water, additives, silica sand) works in the latter case, with chemically hardened moulding material systems one must replace up to 100 % of the curing agents and optional binders with every cycle of the moulding material because they are capable of binding only once. This is similar to an adhesive which also can exert its adhesive effect only once. So reclamation is crucial for all chemically bonded moulding material systems and therefore, also for the inorganic process discussed here. The property of 'reclamation capability' is a feature for the characterization of chemically curing moulding materials and the moulding material process. In order to evaluate the effectiveness of reclamation technologies or systems, the consideration of three characteristics is necessary. This is presented briefly below [3.7].

The reclamation degree α_R describes the percentage reduction in a characteristic test value in the reclaimed sand (reclaimed old sand) when compared to the old sand. Ideally, the test size is recycled back to 0 % in the reclaimed sand, which means a reclamation efficiency of 100 %. The following simple formula describes the reclamation degree α_R:

$$\alpha_R = (P_A - P_E) / P_A \times 100 \%$$

Whereby, P_A is test value in used sand (base state) and P_E is test value in the reclaimed sand (final state).

Typical test values to determine the degree of reclamation are the loss of ignition and the content of fines for organically bonded moulding materials. For example, in the cement moulding material process the CaO content is measured, and in the water glass moulding materials the Na_2O content in the sand is measured. Usually reclamation levels of 100 % are not achieved however, it is desirable to work with values of over 90 %. Under practical use lower reclamation degrees are quite workable.

Reclamation output A describes the relationship from the old sand and reclaimed sand taken from the reclamation plant. The difference between these two masses includes the sand from purified components of binder, silica sand abrasion and third party components, e.g. riser residue. To determine the optimal reclamation deployment one should calculate the sum of the added binder and hardener quantities along with any used additives. To this value the amount of sand abrasion is added which is dependent on the morphological and granulometric characteristics of the sand. The reclamation output should be at about 90 % with many moulding materials reaching values of 95 % or more.

The third characteristic is **Re-use Degree W.** This describes the amount of the reclaimed sand which can be used for moulding material or core making. This amount will depend on the technological requirements of the corresponding moulding material part and is generally between 50 % and 100 %. It is also possible that for specific cores 100 % new sand must be used. For backfill moulding materials 100 % reclaimed sands can be used. Average new sand additions are between 10 % and 30 %.

A fourth characteristic in connection with reclamation is the **Component balance**, which is described by moulding material composition established after an infinite number of reclamation cycles.

A old sand reclamation process includes the following steps:

1. Lump crushing the unpacked old moulding materials.
For this purpose the used sand is first comminuted on the grate basket after the casting, moulding box and larger metal parts have been removed. The degree of crushing is congruent on the mesh size of grate basket. If the grate has a larger mesh, a subsequent crushing unit is also used. If the grate has smaller holes (30 mm to 50 mm in application) this process step is finished. However, attention must be paid to the retention period of the old sand on the grate.

2. Metal separation.
In the second step all metal components must be removed from the used sand. These can be sprayed metal, cooling fins or smaller reinforcement parts, for example. While in iron casting magnetic separators are predominantly used, aluminium foundries use either screening devices or sometimes eddy current separation systems.

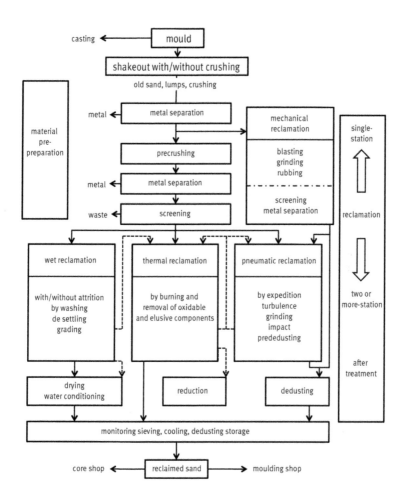

Figure 6.1: Reclamation technologies, classification according to *Weller* [6.1]

3. Reclamation of the used sand.

Next is the actual main step, the removal of old binder shells and components of the sand grains. Principally, the process used for this are cold reclamation (mechanical or pneumatic), thermal or wet reclamation. These processes are described below. All technologies have their limits and applications.

4. Cleaning reclaimed materials.

Reclaimed material is only as good as it's de-dusting. The binder components removed from the sand grains in the previous steps such as powder forming products like silica sand abrasion, feeder or surface residues, etc. must be removed from the system by an effective de-dusting. Otherwise the binder and hardener requirement increases while the gas permeability of the moulding material decreases and risk of gas bubbles in the casting increases. A well-cleaned and dusted reclaimed material (required for use with rounded sand grains) generally has a lower binder and hardener contents than new sand.

5. Cooling of the reclaimed materials.

With increasing temperatures of the environment, and frequent rotation per day, the moulding material often heats to the critical maximum temperatures of about 30 °C. As already mentioned, the manufactured moulding materials and cores change through the process of hardening to a point of being non re-usable. Therefore, it can be useful to cool the material. This can be achieved through appropriate sand coolers or by the addition of corresponding (cold) new sand. Figure 6.1 shows the current recovery technologies for the recycling of foundry sands. These are described in more detail in the following sections.

Mechanical or pneumatic reclamation

The mechanical and pneumatic reclamation processes are often referred to as cold reclamation. This means that these systems are generally operated without additional heating of the used materials or sands. Preheating facilities that are required when working with aqueous materials to rejuvenate dry, old sand are excluded. Mechanical reclamation has some advantages. There are many systems on the market which can be used in principle for all moulding material systems (organic or inorganic). They are attractively priced and have low operating costs. Most are also modular and can be adapted to various used sand operations.

Mechanical recovery systems work on the principle of blasting, grinding or rubbing. This means that cured binder shells are removed by rubbing of the grains against one another, against the friction tool, or against the container walls. Examples of such systems are Ball mills, fluid bed cleaners or abrasive drums. They can be used either as the complete reclamation plant, but can work also as the first reclamation stage for a subsequent pneumatic reclamation system. Simple devices such as screen filter sets or spiral grain crushers are separating systems and cannot be used for the reclamation of used sands.

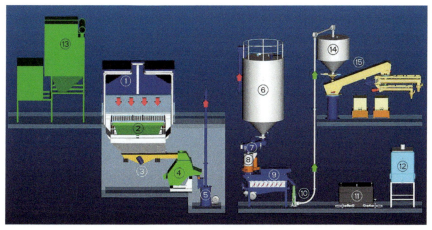

Figure 6.2: Layout of a mechanical reclamation system (Fig. AAGM / Woehr, Bopfingen)

Picture explanation:
1 – dust and noise control cabin
2 – shakeout
3 – vibrating conveyor trough
4 – crushers
5 – valve lift
6 – hot sands silo
7 – separating screen
8 – centrifugal cleaner (reclamation unit)
9 – fluid bed cooler classifier
10 – pneumatic conveyors
11 – cooling water tank
12 – water cooling tower
13 – filter
14 – reclaimed sand silo
15 – mixer

The pneumatic reclamation of old sand is by air acceleration and deceleration. The energy existing in the sand is converted when decelerating into reclamation effect and ensures the removal of old binder constituents in the sand grains. In the early days of this technology, the sand was bounced against an obstacle during acceleration, for example, against an impact bell. However, in this procedure there is an increased grain wear because the impact may break some of the silica sand grains. Newer systems slow down the moulding material before striking so the cleaning of the sand grains occurs during movement and in the moment of braking. Figure 6.2 shows a schematic layout for a hand moulding material with mechanical reclamation. The vast majority of today's reclamation plants are mechanical or pneumatic.

Thermal reclamation

Thermal reclamation is suitable for organically bonded used sands. The principle of these systems is based on the combustion of the old binder constituents. The working temperatures are between 600°C and 900°C, and typical systems are rotary kilns or fluidized bed furnaces. The free accessibility of the entire sand surface for combustion is important in this reclamation technology. In the fluidized bed furnace, for example, this is guaranteed. Thermal reclamation systems produce high-quality reclaimed sands and production process residues such as dusts are inert and can be removed easily. The main disadvantages of thermal plants are their enormous size and their high-energy requirements. This means that an economical operation of a thermal reclamation plant is guaranteed only with bulk amounts and continuous use. Because of this, the number of operating thermal plants is quite low.

Wet reclamation

Wet reclamation of used sand is only feasible for water soluble binder systems such as those based on alkali silicates and water glass. Wet reclamation is based on the principle of attrition; the used sand is cleansed through movement. The result of this technology is a reclaimed sand type that is very similar to new sand and has excellent features. Disadvantages are the space requirements (since the systems are usually constructed as a cascade scrubber), water consumption of the plant and wear associated with the sewage treatment. If the process water is purified, sludge remains as residue which poses problems for disposal under certain circumstances. Although this process is often discussed in the field of inorganic binder systems, in Germany there are no wet reclamations systems in operation.

Reclamation of inorganically bonded moulding materials

Reclamation of water glass used sands

Efforts to reclaim used sands from the inorganically bonded moulding materials and reuse them for moulding material and core production appeared quite quickly after the introduction of the water glass CO_2 process as publications such as *Klat* [6.2], *Palmer* [6.3] and *Roberts* [6.4] prove. According to *Flemming* [6.5], out of the variety of existing publications, only some contributions are worth mention. In [6.6] *Jennes* describes a relative reclamation degree in addition to an absolute reclamation degree. This is defined by a certain initial state of the used moulding material to an absolute degree of attained reclamation compared to a certain degree of reclamation required for application (figure 6.3). It is of importance that there is no linear relationship between the in-

crease of the (absolute) and the degree of reclamation of expended reclamation work. This relative degree of reclamation is therefore, dependent on three variables:
1. The concrete production technology (moulding material – casting).
2. The economic, qualitative and quantitative limits due to the mechanical design of the system components.
3. The desired or necessary absolute degree of reclamation.

From the point of view of reclaimed sand quality, the best variant for reprocessing of water glass used sand is without a doubt, wet reclamation. By incorporating a water soluble binder system, it is possible to free the resulting used moulding materials by 'washing' of the old binder shells which can result in the reclaimed sands reaching or even exceeding new sand quality. Reclamation plants that operate on the basis of attrition are mentioned among others in [3.7]. Such wet reclamation equipment in the Czech Republic was at one time widespread although today it has become rare. The reclaimed sand quality is very good even when considering the high water demand, increased maintenance of the equipment (water tight!) and required extensive cleaning of the resulting wastewater. Although there have been some attempts at wet reclamation in recent years, it has once again been excluded from mainstream use.

In times of greater dissemination of water glass moulding materials and procedures, a wide variety of systems were used to reclaim old sands. *Röhrig* in [6.7] uses the example of a fine lump impact crusher which operates on the principle of impact grinding. Because here, used core sands from the water glass process of CO_2 are to be reclaimed and the importance of the separation of bentonite bonded circulation moulding material from used core sand is pointed out. Furthermore, the author reports that improving the strength of the form mixtures is possible by using 50 % reclaimed sand. This is due to the rounding of the grains from the stress of the reclamation.

Although the reduction of the sintering start (1500 °C new sand, reclaimed sand 1400 °C) is moderate, usage is discouraged in the production of steel castings. As a parameter to monitor the reclaimed sand, *Röhrig* proposes the determination of the content of fines and ignition loss. Although the impact crusher is suitable for the reclamation of water glass used sand, there has been a turn more recently to newer and gentler reclamation technologies.

The comparison between thermal and mechanical pneumatic reclamation is reported by *Flemming inter alia*, in [6.8]. The thermal plant uses natural gas which makes necessary, from an economic point of view, the heat utilization from the exhaust gases. In the mechanical pneumatic reclamation system, a mixture of 70 % furan resin used

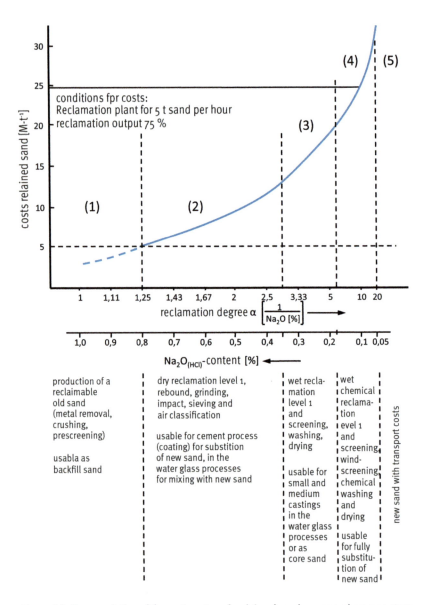

Figure 6.3: Representation of the cost per ton of reclaimed sands, uses and process steps, depending on the degree of reclamation, *Jennes* [6.6]

sand and 30 % used water glass sand is reclaimed. The reclamation produced in a five cell pneumatic reclamation system can be used again to 50 % in the furan resin process. It is interesting that 30 % of screened (not reclaimed) used sand can be successfully used for the production of water glass ester moulding material. A thermal reclamation plant working at 800 °C can be operated according to this economically appropriate utilization. In [6.9], *Vernay* also describes the reclamation of water glass ester moulding materials and arrives at maximum re-usage efficiency of 88 %. To check the properties of moulding material *Vernay* uses parameters of moisture variables, loss of ignition, soda residue, particle size distribution, processing time and compression strength. *Chahan* and *Cobett* use a pneumatic system in which the sand grains are mixed in a rotary motion thereby preventing the electrostatic adhesion of the binder residue from the grains [6.10]. Therefore, the recovery level of the reclaimed sand is 80 % given when 0.25 % to 0.75 % water is added to the moulding material. Without this water, the moulding material mixture reacts very quickly, which has the effect of low final strength. The authors use 4 % water glass binder (Module 2.6) and 0.4 % ester hardener. Reclaimed water glass sand can thus be used even for the production of 50 % oil bonded moulding material, wherein also water is added.

In [6.11] *Lissell* poses the question of whether or not, in dry or wet reclamation, water glass bonded used sands should be used as a preference. With regard to the wet reclamation, he noted that the water solubility of the silicate binder films decreases with increasing temperature. From this he infers that only used sand with a low thermal load (maximum 400 °C) can be reclaimed economically (figure 6.4). In his practical studies on the reclamation of chromite sands he also applies less effective dry reclamation process and uses a two-cell pneumatic system. The content of Na_2O serves as an indicator for the determination of reclamation efficiency. He compares different process for the determination of the sodium content in the investigated moulding materials. The results are shown in table 6.1. This shows that a complete solution of the samples is not possible, that solution in various acids gives comparable results, and that boiling in water is unsuitable, and that results from being mixed into cold water with simultaneous titration closely compare with the acid application process. In the described tests 50 % of the water glass waste was recovered during the crushing and screening steps, while in the actual reclamation stage about an additional 25 % was realized. In total, 75 % of the used sand can be recovered. Table 6.2 shows the Na_2O reductions that could be achieved by the reclamation on different days. The water glass binder used had a modulus of 2.7 and contained a sugary disintegration aid which was added at levels of between 3 % and 4 %. Diacetin and triacetin combinations were used as a curing agent.

In [6.12], *Tilch* among others discuss the question of the applicability of the various reclamation technologies employed in foundry moulding material systems. Figure 6.5 also gives a corresponding representation. The practical investigations are carried out with the following process variants using an airflow grinding plant:
- water glass ester moulding material
- water glass clay molasses moulding material
- water glass cement moulding material

Polzin, Nitsch and others describe studies on the reclamation of water glass and bentonite used sands. These operated on the principle of the fluidized scouring bed reclamation system [6.13]. This is a purely mechanical flow bed cleaner which is equipped with a powerful dust collection system. Four different sands were reclaimed from the water glass process whereby material mixtures comprised of 100 % used sands reached a recycled strength of 78 %–80 % compared to new sands. The four investigated water glass ester used sands were not thermally treated however, the process of the fluidized bed air cleaner was applied at a temperature of 100 °C to remove any remaining water contained in the moulding material. Moisture in the waste sand leads to the formation of a tough plastic used binder cover that it is very difficult to remove by mechanical reclamation. These studies have shown that the fluidized bed cleaner is basically suited for the reclamation of water glass used sands. In production with moulding material mixtures containing reclaimed sands, a 1 % addition of water is used. In [6.15], these studies are supplemented inter alia by evaluating the penetration tendency, the Na_2O contents and the initial sintering temperatures. As illustrated by table 6.4, the Na_2O values compared to the old sand can be significantly reduced and achieve a maximum of 0.15 %. These results are essentially consistent with the opinion of other authors who state that the Na_2O values should not exceed the maximum value of 0.2 %, at least in the area of iron casting. If this percentage is exceeded it will result in a significantly worsening sintering behaviour of the processed material. The considerable

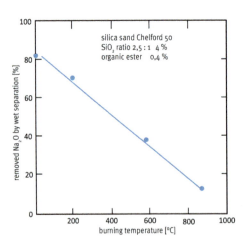

Figure 6.4: Solubility of water glass binder shells in water according to the thermal loading of the moulding material [6.11]

Table 6.1: Comparison of different process for the determination of Na_2O content in moulding materials [6.11]

Sand	Extraction process	Investigation process Na_2O	Content Na_2O [%]
Olivine sand, Water glass CO_2 curing used sand	Cock in 5 % HNO_3	Atomic absorption	0,335
	Complete solution	Atomic absorption	0,330
Chromite sand, Water glass ester curing Reclaimed sand	Cock in 5 % HNO_3	Atomic absorption	0,325
	Complete solution	Atomic absorption	0,320
Olivine sand, Water glass ester curing Reclaimed sand	Cook in 5 % HNO_3	Atomic absorption	0,47
	Complete solution	Atomic absorption	0,39
Chromite sand, Water glass ester curing			
I From mixer	Cock in 5 % HNO_3	Atomic absorption	0,39
II From mixer			0,30
III Crushed			0,23
IV Reclaimed			0,11
I From mixer	Wet scrubber	Titration	0,40
II From mixer			0,35
III Crushed			0,23
IV Reclaimed			0,07

Table 6.2: Na_2O content in used sand, broken used sand and pneumatically reclaimed moulding material

	Sand from mixer	Crushed and screened used sand		Pneumatically reclaimed sand		
No.	Ø Na_2O-content [%]	Ø Na_2O-content [%]	Na_2O removed [%]	Ø Na_2O-content [%]	Na_2O removed [%]	Removed totally Na_2O [%]
I	0,517	0,231	55,4	0,124	46,3	76,1
II	0,479	0,181	62,2	0,124	31,5	74,2
III	0,431	0,181	58,0	0,112	38,2	74,0
IV	0,363	0,167	54,0	0,112	33,0	70,0
Average	0,448	0,190	57,6	0,118	37,9	73,7

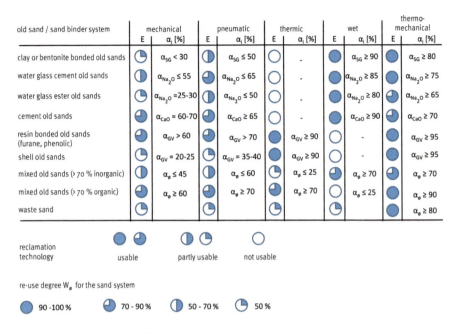

Figure 6.5: Applicability of different reclamation processes [6.12]

values for the initial sintering temperatures in table 6.4 are a further indication that the reclamation unit used here can be used for water glass materials. The remarkable values for the initial sintering temperatures in table 6.4 are a further indication that reclamation aggregate used here can be used for water glass moulding materials.

Similar tests were carried out by *Döpp* inter alia. Here the authors investigate the water glass CO_2 and the water glass bentonite processs in addition to the water glass ester process [6.16]. The tables 6.5 and 6.6 show characteristics of the studied material and results which clearly demonstrate the suitability of the reclamation aggregates used. An additional interesting result of this work is the decay behaviour shown in figure 6.7 based on the residual compression strength. Thus, the addition of between 0.7 % and 1.0 % bentonite causes very low residual strength. Even with an increasing number of reclamation cycles there is no significant deterioration.

In a much publicized project in 1998 [6.17], a central reclamation facility was installed in the Ernst Bröer Aluminium Foundry in Schwetzingen. This was for the purpose of the reclamation of water glass CO_2 used core sands from four different aluminium foundries and the reinstating of these sands to the original application in the core production. Through this approach, 1500 tonnes of used sand was gathered annually and was sufficient for the economic operation of the plant. As the reclamation plant, the 'Jet Reclaimer' served as a pneumatically working unit wherein the used sands were dried at 200 °C with hot air and accelerated in an ascending pipe while being conveyed

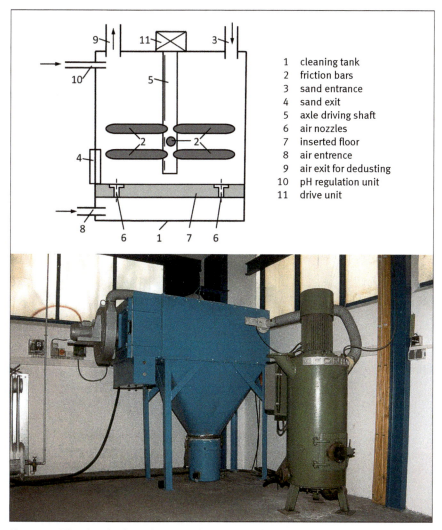

Figure 6.6: Pilot plant for the reclamation of inorganically bonded used sands [6.13], [6.14]

Table 6.3: Data and characteristics of reclaimed water glass used sands [6.13]

	Used sand A	Used sand B	Used sand C	Used sand D
Results oft he reclamation investigations				
Binder	DESIL®	DORSIL® 450	ANTEF® C4	ANTEF® M1
Casting alloy	Cast iron	Cast iron	Cast iron	Al Bronze
Binder content [%]	3,0	3,0	3,0	3,9
Hardener content [%]	0,3	0,3	0,3	0,39
AGS new sand [mm]	0,31	0,31	0,31	0,33
Dust amount and reclamation values of water glass old sands				
Dust amount [%]	10,4	7,7	7,7	1,9
pH value dust	10,35	10,47	10,90	11,94
LOI [%]	6,7	6,12	7,1	14,7
el. conductivity [µm/cm]	9405	7800	7570	773
Results oft the reclamation investigation				
Used sand from...	Iron foundry			Copper foundry
Binder	ANTEF® C4			ANTEF® M1
Binder content [%]	3,0	3,0	3,0	3,9
Hardener	ANTEF® K1			ANTEF® K3
Hardener content [%]	0,3	0,3	0,3	0,39
Water [%]	1,0	1,0	1,0	1,0
Reclamation time [min]	30	30	30	10
Bending strength new sand (24 h) [N/cm²]	108	108	108	107
Bending strength reclaim. sand (24 h) [N/cm²]	83	87	92	102
Bending strength reclaim. sand vs. new sand [%]	78	80	86	96

Table 6.4: Na_2O contents and initial sintering temperatures of four water glass ester used sands and their reclaimed sands [6.15]

	New sand		Used sand		Reclaimed sand	
	content Na_2O [%]	Sintering start point [°C]	content Na_2O [%]	Sintering start point [°C]	content Na_2O [%]	Sintering start point [°C]
A	0,001	1250	0,46	700	0,060	1100
B	0,001	1250	0,44	700	0,072	1150
C	0,001	1250	0,31	750	0,063	1250
D	0,007	1150	0,46	550	0,150	900

Table 6.5: Characteristics of the studied water glass materials [6.16]

Moulding material	Binder		Hardener	Additive
	Water glass ester process Water glass SOLOSIL 2000		Ester VELOSET 1	Water
New sand H 32	3 %		0,3 %	-
Used sand*)	2 %		0,1 %	0,7 %
	Water glass CO_2 process Water glass		CO_2	Water
New sand	3 %		2 bar 20 s	
Used sand	Mostly 3 %		2 bar 20 s	0,1 – 0,3 %
	Water glass bentonite ester process Water glass	Bentonite	Ester	Water
New sand	a) 3 % b) III c) 30 s	1 % I 20 s	0,3 % IV 45 s	0,2 % II 30 s
Used sand	a) 2 % b) III 30 s	0,7 % II 30s	0,1 % IV 45 s	0,2 % I 30s
	Water glass bentonite CO_2 process Water glass	Bentonite	CO_2	Water
New sand	a) 3 % b) III c) 45 s	1 % I 20 s	2 bar 20 s	0,1 % II 30 s
Used sand	a) 3 % b) III c) 45 s	1 % II 30 s	2 bar 20 s	0,1 % I 30 s

*) Used sand – 100 % reclaimed sand, from 2nd cycle
a) Content, b) addition sequence, c) mixing time or gassing time

Table 6.6: Results of the mechanical reclamation with the fluidized bed cleaner [6.16]

Water glass process (hardener)	Sand[1]	AGS [mm]	Na$_2$O Content [%]	Compression Strength[2] [N/cm^2]	Compression Strength[3] [N/cm^2]	Bench life[4] [min]	Residual Strength[5] [N/cm^2]
(1) Ester	New sand	0,234	-	-	396	10	210
	Reclaimed sand	0,218-0,231	0,1-0,14	-	402-460	9-12	160-350
(2) CO$_2$	New sand	0,230	-	29	314	72	340
	Reclaimed sand	0,215-0,227	0,11-0,15	33-46	315-360	65-79	283-490
(3) Bentonite Ester	New sand	0,234	-	-	302	10	63
	Reclaimed sand	0,217-0,23	0,06-0,11	-	308-422	9-11	20-36
(4) Bentonite CO$_2$	New sand	0,234	-	43	254	52	85
	Reclaimed sand	0,216-0,23	0,09-0,13	41-51	308-350	45-53	32-67

[1] New sand H 32, reclaimed sand from 1nd to 8nd cycle; [2] compression strength directly 30 s; [3] compression strength after 2 days storage; [4] storage time of the sand mix with the strength after 2 days of 70 % of the start value; [5] 30 min at 800 deg C, measurement after 1 day storage

upward (figure 6.8). This movement created an intense grain-to-grain friction through which old binder covers are removed. Unlike previously used reclamation plants with a similar design and baffle plate for the sand, the sands in the Jet Reclaimer were stopped against the tubing located above the baffle plate before impact. Through this a more intensive separation was achieved from the binder covers and significantly reduced grain wear resulting from broken grains. Unfortunately, this promising project remained an isolated case and was discontinued some time later. The central reclamation facility was frequently the subject of mind games and project ideas for many foundries for several years. Because of all the economic arguments for such a solution one must hold it in discussion. However, many questions remain such as transport routes and costs, different plant design parameters for various types of used sands and the strict separation of the individual moulding materials in such facilities.

The reclamation of water glass used sands exists in several approaches in practical operation. This helps in assessing the current behaviour of these form substances. [6.18] describes such work. The fluidized bed cleaner, which has been described here several times, is used for water glass ester reclaimed sand moulding materials. After the casting of the test specimens with aluminium bronze alloy, the next reclamation follows with the recast test geometry which contains 100 % reclaimed sands. Bending

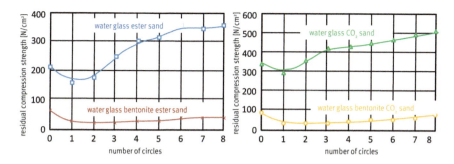

Figure 6.7: Residual compression strengths of various water glass moulding materials with and without bentonite additives (left water glass ester, right water glass CO_2) [6.16]

strength serves as the command variable for the evaluation of recycle behaviour. Figure 6.9 shows that the strength increases already after the first round of reclamation by approximately 30 % and then up to 5 % thereafter. It stays at this level during subsequent cycles. Thus, it can be demonstrated that there is actually an improvement in the properties of moulding material through the dedusting and rounding of the grains. However, the continued use of 100 % reclaimed sands (virtually impossible over the long term) will result in a deterioration of other moulding material characteristics.

Starting at a cycles of 900 °C these problems can be reduced by the addition of new sand at operational norms of 10 % to 30 %. The high efficiency of the equipment used for the separation of the used sand grains from old binder shells is shown in figure 9.10.

In [6.19] an attempt to evaluate (mechanical) reclamation systems in comparative studies is shown. Two of the three reclamation

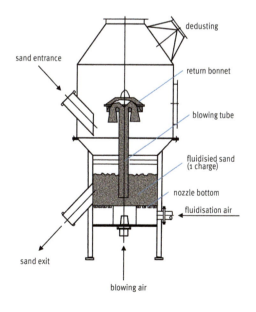

Figure 6.8: mechanical pneumatic reclamation unit "Jet Reclaimer", schematic diagram [6.17]

Reclamation of Used Sands | 179

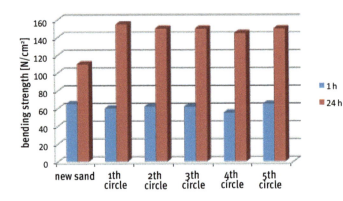

Figure 6.9: Bending strength of moulding material mixtures from 100 % reclaimed sand at five reclamation cycles [6.18]

plants have already been discussed here the fluidized bed cleaner and the Jet Reclaimer. The third is the grinding regenerator, which is described in detail by *Müller-Späth* in [6.20]. The grinding regenerator is a system which is based on the principle of grinding. Three used sands were examined in the context of this work: Two water glass ester sands and one water glass CO_2 sand. Table 6.7 illustrates the characteristics of these used sands. In a slanted, rotating mixer vessel, a countervailing movement of grinding wheels in the moulding material takes place, separating the binder shells from the grains of sand. The grinding regenerator cleans the moulding material intensely and results in reclamation times of 5 to 10 minutes. Results of the comprehensive studies are summarized in table 6.8. Possible areas of use of the used reclamation aggregate can be derived from this data. The achieved recovery levels for the three used sands in the three different reclamation plants are shown in figure 6.13. Images 6.14 and 6.15 show the approach and the result of a practice test with reclaimed sand A. Here, the reclaimed sands from the Jet Reclaimer (reclaimed sand 1) and the grinding regenerator (reclaimed sand 2) were compared with new sand. The produced displacement propeller wing shows a good, uniform surface. These results are underpinned by strength measurements carried out in parallel and demonstrate the suitability of the reclaimed sands to produce fair quality castings. Similar field trials were carried out with the other two used sands.

In addition to [6.19] the evaluation of the cyclical behaviour of the three discussed used sands are studied in [6.22]. Only the fluidized bed was used for reclamation. At five reclamation cycles new sand addition/exchange should be adjusted to 20 % in the moulding material. Table 6.9 shows the experimental conditions and in table 6.10 the results are presented. In summary, it is shown that the repeated reclamation and re-application of water glass used sands in the production of moulding materials and

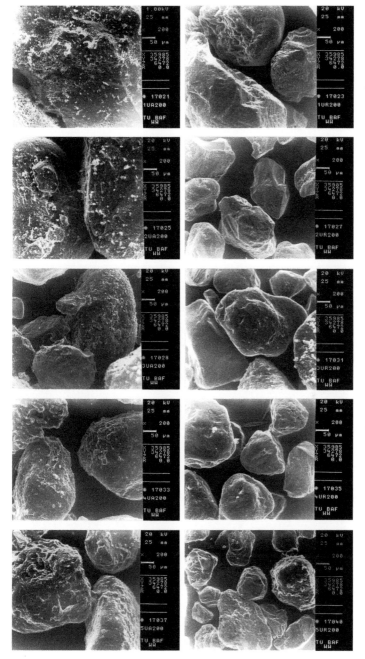

Figure 6.10: SEM images of used sand (left) and reclaimed sands (right) of 1 (top) to 5 (below) reclamation cycles [6.18]

Table 6.7: Comparison of three reclamation plants and examined used sands [6.19]

Used sand A	Used sand B	Used sand C
Water glass ester	Water glass CO_2	Water glas ester
Silica sand	Silica sand	Silica sand
Casting alloy Al Bronze	Casting alloy cast iron	Casting alloy alloyed cast iron
Pouring temperature 1150 °C	Pouring temperature 1350–1450 °C	Pouring temperature 1400–1500 °C
Mono system	Mixed system with with bentonite Cold box- and shell sands	Mono system

Figure 6.11: Jet Reclaimer on ÖGI Leoben, Principle Figure 6.8

cores is possible. The results also show that there is no 'universal' reclamation technology for all used sands. Here, the reclamation of used sand A fulfilled all requirements. Documentation for these results are provided in figure 6.16, whereby an increase in strength emerges from an increasing number of reclamation cycles. This speaks for an effective and gentle removal of the spent binder components in this plant.

Proof of the interest in the problem areas of the reclamation of used sands in the environmentally friendly water glass CO_2 process are shown in the contributions of *Fan and others* in [6.23] and [6.24]. Here dry and wet reclamation are compared in extensive studies. For dry reclamation, the authors determine a reduction of used binders of between 10 % and 20 %, and work with new sand allowances of 25 % to 35 %. When using wet reclamation, the levels of binder residues are reduced by 85 % to 95 %, which is reflected in necessary new sand additions of 5 % to 15 %. As the most economical option, the authors suggest a combination of dry and wet reclamation. The reclaimed sand obtained from this can be used as the sole moulding material. The so-called 'Noram Process' for the mechanical reclamation of used sands is presented by *Genzler* in [6.25]. In this system, the broken and scattered used sands are separated during recla-

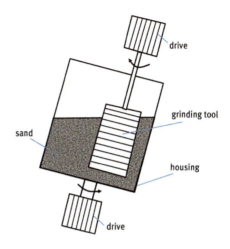

Figure 6.12: Intensive grinding regenerator, mode of action [6.19]

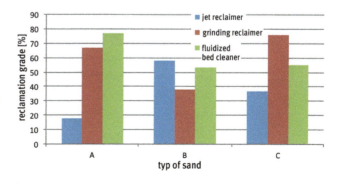

Figure 6.13: Reclamation degrees of the three examined used sands based on the Na_2O content [6.21]

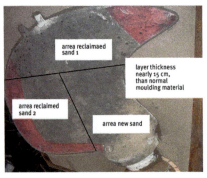

Figure 6.14: Arrangement of reclaimed sands to the pattern [6.19]

Figure 6.15: Casting result for the surface evaluation of adjusting propeller wings [6.19]

Table 6.8: Comparison of three reclamation plants, results, used sand from table 6.7 [6.19]

Parameter	Setpoint	Jet reclaimer	Grinding regenerator	Fluidized bed cleaner
Bending strength 100 % Reclaim sand	80 % New sand value	A (350 °C), C (75 %)	A (76 %), B, C	A (110 °C), C (110 °C)
Sintering start	min. 800 °C	B (A,C)	A,B,C	A,B,C
El. conductivity	Conductivity⇩, Sintering point⇧, Strength⇧, Correlation to Na$_2$O content			
Practical test strength	Reclaimed/ new sand 50/50 und 70/30	A,B,C, B 100/0	A,C, B n. b.	not measured
Practical test surface	Reclaimed/ new sand 50/50 and 70/30	A,B,C, B 100/0	A,C, B n. b.	not measured
Advantages	–	Grain friendly, no moving parts in sand	Short reclamation times, better for mixed sands	Good- reclamation, grain friendly
Disadvantages	–	Longer reclamation times	Moving parts in sand, high wear	vulnerable- to wear

Table 6.9: Characteristics of tests for circulating behaviour of water glass used sands [6.22]

	Used sand A	Used sand B	Used sand C
Silica sand AGS	0,51 mm	0,24 mm	0,27 mm
Binder 3,5 %	S 26	P 58	C5
Hardener 0,35 %	K5	CO_2	K3
Mould material/Core	Mould material/Core	Core	Mould material
Reclaim sand amount	350 kg	15 kg	30 kg
Casting alloy	CuAl10Ni	GJS-500	Alloyed ducile iron
Pouring temperature	1140 °C	1450 °C	1350 °C
Casting	Propeller blade	Housing	Cylinder
Mass casting	30 kg	18 kg	12 kg

Table 6.10: Results of the tests for cycle behaviour [6.22]

	Requirement	Used sand A	Used sand B	Used sand C
Bending strength 100*	80 % New sand	Yes	Yes	No
Bending strength 80/20**	80 % New sand	Yes	Yes	(Yes)
Sintering start 100	Min. 800 °C	No (700 °C)	No (700 °C)	No (700 °C)
Sintering start 80/20	Min. 800 °C	Yes	No (700 °C)	No (700 °C)
Conductivity 100	increase/decrease	increase	decrease	decrease
Conductivity 80/20	increase/decrease	decrease	increase	decrease
Na_2O content 100	0,1 – 0,2 %	Yes	No	No

* 100 % Reclaimed sand, ** 80 % Reclaimed sand, 20 % New sand

mation according to oversize and undersize with a counter-current classifier. Thereafter, the actual reclamation is carried out in an attrition mill. The reclamation performance of the system, which is controllable via the power supply, can ensure optimal results with minimal equipment wear.

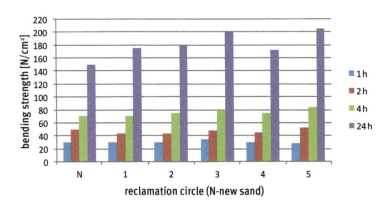

Figure 6.16: Bending strengths reclaimed sand A – 100 % reclaim (curing time in legend) [6.21]

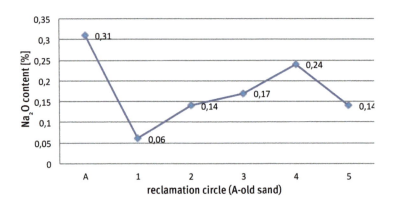

Figure 6.17: Na_2O contents used sand A – 100 % reclaimed sand in five cycles [6.22]

Danko reports in [6.26] on property improvements of the reclaimed materials produced at lower application temperatures during reclamation (figure 6.18). Reclamation at temperatures of -70 °C results in stress formations such as increased brittleness in the binder bridges. The process can be applied in principle to other water based moulding material systems, i.e. also for bentonite bonded moulding materials.

Although the reclamation of water glass used sands is illustrated by as many examples as possible, potential users are often still skeptical. Often the question is posed concerning reference plants which, at least in Germany, do not exist at the moment. The

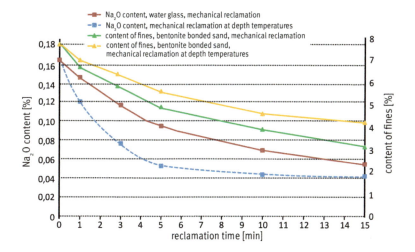

Figure 6.18: Na_2O content of water glass sand and content of fines for bentonite bonded moulding materials in reclamation in ambient conditions and low temperatures [6.26]

trial that should close this gap is presented in [6.27]. The paper presents results of an eight month test phase working with a mechanical reclamation plant for water glass ester used sands at room temperature in a medium sized foundry. This pilot test was to determine the degree of reclamation of the plant and for setting the maximum usable replenishment quantities for the production of hand moulded aluminium castings. This plant used an intensive grinding regenerator. The system basically consists of a crusher, the grinding regenerator itself, a separator and a filter system.

In the preliminary experiments, the first step was to establish the optimal system parameters for the grinding reclaimer. At a grinding time of 5 minutes, a low dust discharge paired with a good Na_2O content was realized along with a comparable initial bending strength to what is achieved with the use of new sands. A short grinding time of 5 minutes ensures a fast process flow. Because these settings initially did not have the desired success, some changes were made to the test procedure. The circulation rate was reduced from 5 tonnes to one tonne of moulding material. The ratio of used sand to new sand in the moulding material is kept at 90 : 10 % because good results in previous tests were obtained at that figure. The binder content was determined to be optimal at 2.5 % as in the first series of experiments. Two different reactive hardeners were applied at a dose of 0.25 %. Seven minutes of grinding with grinding wheels of

Figure 6.19: Test facility with intensive grinding reclaimer, separators and filters (breaker outside the hall) [6.27]

400 mm diameter and high gear ratio were set as the setting for the grinding regenerator. An aluminium part (figure 6.21) was selected as the manufactured casting with the following parameters: AlSi7Mg alloy with a casting weight of 66.1 kg and a 663 kg form weight.

The study confirmed once again that this mechanical reclamation system is basically suitable for the reclamation of used sands from the water glass ester process. In seven reclamation cycles, the statistical exchange of moulding materials was simulated successfully without negative effects on the quality of the cast part. It was found that a new sand addition of 10 % is sufficient in principle. The reclamation effect could be demonstrated inter alia on the basis of the Na_2O content while the reclamation levels however, were partial only at 50 % (figure 6.22). As in previous comparable studies, it was determined that in the area of aluminium castings, higher levels of Na_2O are allowable while in the area of iron castings the levels should be in the range of approximately 0.2 % to 0.5 %. The measurement of the electrical conductivity is suitable for the quick determination of the reclamation effect (approximately 700 µs/cm) (figure 6.23).

For the success of reclamation the moulding material moisture is of crucial importance, and in this specific case the setting was set at a maximum of 0.1 %. The initial sintering temperatures of the reclaimed sands were on average only about 650 °C which, in the production of aluminium castings, is a manageable problem. During reclamation the binder content can be significantly reduced due to the rounding of the grains. It should be noted that required improvements in the pilot plan included dust removal. Further refinements of the system parameters are expected to produce even better results. For applications such as in the area of iron castings, for example, the necessary reclamation parameters are yet to be determined.

Usually the process of determining the reclamation levels is ascertaining the Na_2O content. Another possibility for quality control of regenerated moulding materials is described by *Pavlowsky inter alia,* in [6.28] which is the use of capillary Isotachophoresis to determine the levels of sodium acetate in the moulding material. The Isotachophoresis (ITP) is an analytical procedure used, for example, in the testing of water samples. The ions to be analyzed are separated using an electrolyte solution in an electric field. This happens due to different ion mobilities. Through the use of a discontinuous buffer system with a counter ion, the different ions move at the same speed after the separation.

Figure 6.20: Open reclaimer with abrasive grinding tool [6.27]

Figure 6.21: Aluminum casting produced in the reclamation experiments [6.27]

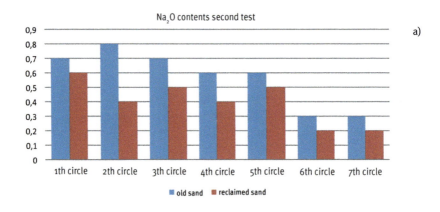

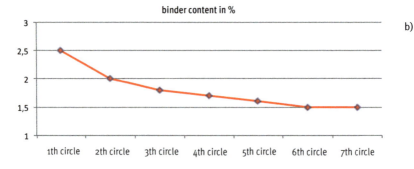

Figure 6.22: Na$_2$O content of used and reclaimed sand (a) and binder contents (b) [6.27]

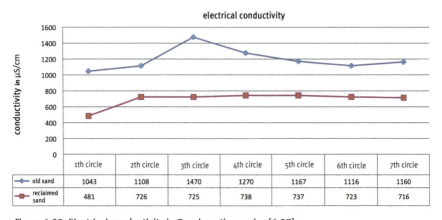

Figure 6.23: Electrical conductivity in 7 reclamation cycles [6.27]

An electrolyte is used which has a high ion mobility and thus moves quickly in the field. It is in front of the sample in the separation capillary and is therefore, called the Leading Electrolyte. An electrolyte having a low ion mobility and migrates slowly through the field is located behind the specimen in the separation capillary, and is therefore, called Terminating Electrolyte. Because the field between these two electrolytes has special properties, all ions between them receive a particular order and concentration and are also independent from the initial state. This procedure is similar to chromatography.

For synthetic and operating tests, the results of these investigations demonstrate that the measurements show very good matches with the existing acetate contents. A commonly used limit for sodium acetate is a maximum of 0.15 %, while content in reclaimed materials are often between 0.25 % and 0.40 % acetate. The measurement process has a lower limit of determination of approximately 0.023 mass % and can detect acetate levels upwards of up to 1.8 %. Using these figures, a foundry in the Czech Republic applies the process described for assessing reclamation levels.

Reclamation of water glass bonded alumino-silicate sands (mullite) is the subject of [6.29]. The tables 6.11 and 6.12 contain the property values of the used mullite raw moulding materials as well as the output date of the new sand mixture. The reclamation unit used in these tests was the aforementioned fluidized bed cleaner, and the cy-

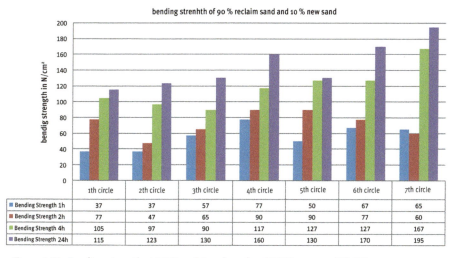

Figure 6.24: Bending strength at 90 % reclaimed sand and 10 % new sand [6.27]

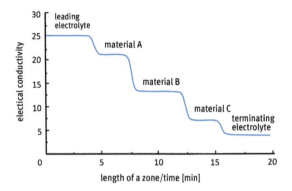

Figure 6.25: Example of a isotachophoresis curve for different cations [6.28]

cle time was set at thirty minutes. Based on the bending strength (see table 6.13) it can be concluded that the reclaimed mullite sand can be re-used for making moulding materials. The start of sintering here decreases to about 600 °C and remains relatively constant in subsequent reclamation cycles. Unlike silica sand, the electrical conductivity settles between 1200 µS/cm and 1400 µS/cm, while the Na_2O content rises up to 0.9 % in the fifth round. This may be related to both the process parameters of the small scale equipment used as well as the surface structure of the mullite. If this leads to difficulties during the preparation of the reclaimed sand the response should be the addition of more new sand. In these described works the moulds were cast with aluminium alloys. Lesser numbers of experiments were made with steel alloys.

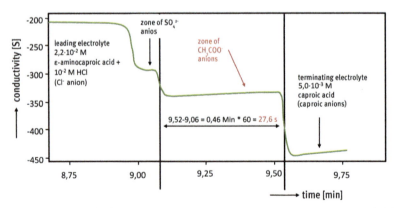

Figure 6.26: Isotachophoresis curve of a sample with 0.1 wt% $Na_2CO_3 * H_2O$, cured with a 50/50 mixture of diacetin and triacetin [6.28]

Reclamation of used cement sands

The 1940s and 1950s were the heyday of the application of the cement form process. During this time the reclamation of used sands was not at the forefront of thought within the industry. However, publications on the subject of reclamation of cement mould-

Table 6.11: Characteristics of the used mullite [6.29]

Parameter	Value
Density g/cm^3	2,9
Bulk density g/cm^3	1,6
Sintering start beginn (VDG) °C	1250
Fire resistance heating microscope °C	> 1740
Linear expansion coeff. α (20-600 °C) 10^{-1}K^{-1}	4,0
AFS-Number	65+/- 5
Grain shape	Spherical

Table 6.12: Moulding material parameters new sand mixture with mullite (2.5 % sodium silicate, 0.25 % hardener) [6.29]

Test parameters	Results
Content of fines [%]	0,02
Sintering start [°C]	1200
Electrical conductivity [µS/cm]	0,04
Bending strength, 1 h [N/cm^2]	35
Bending strength 2 h [N/cm^2]	55
Bending strength 4 h [N/cm^2]	95
Bending strength 24 h [N/cm^2]	230
Na$_2$O-content [%]	0,5

Table 6.13: Comparison bending strengths of mullite moulding material mixture (10 % new sand) at 5 cycles [6.29]

σ_B [N/cm^2]	New sand	Cycle 1	Cycle 2	Cycle 3	Cycle 4	Cycle 5
after 1 h	35	47	30	32	42	32
after 2 h	55	63	65	65	57	62
after 4 h	95	105	107	107	97	103
after 24 h	230	212	215	247	232	237

ing materials can be found from this era. For example, *Schmidt* looks at the reclamation of a number of materials including sands from the then popular cement molasses moulding process [6.30]. Pneumatic dry reclamation was the applied process. Figure 6.29 shows the variations of grouting material, CaO and C content, and the loss of ignition in the mould from repeated reclamation of these substances. Quality castings of

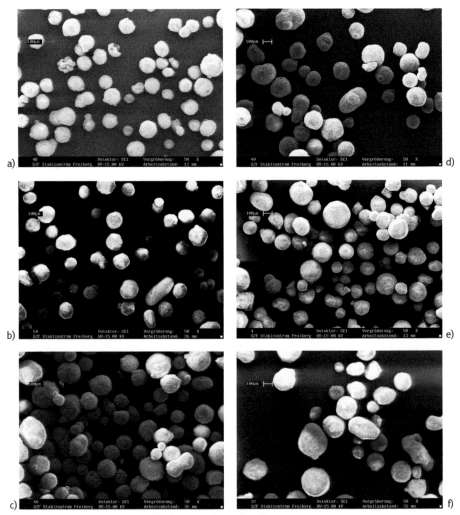

Figure 6.27: SEM images, particle shape, reclaimed mullite
(new moulding material 1–5th reclaimed sand b–f) [6.29]

1.8 tonnes of mass were made here with 100 % reclaimed materials. Since the pollutant content had not yet reached the state of equilibrium in the third cycle, the permissible continuous reclaimed sand was 80 %. The possibility of wet reclamation n for the cement moulding material process was also discussed by various authors at this time. *Zimnawoda* provides the basic structure of such an installation in figure 6.30. Central

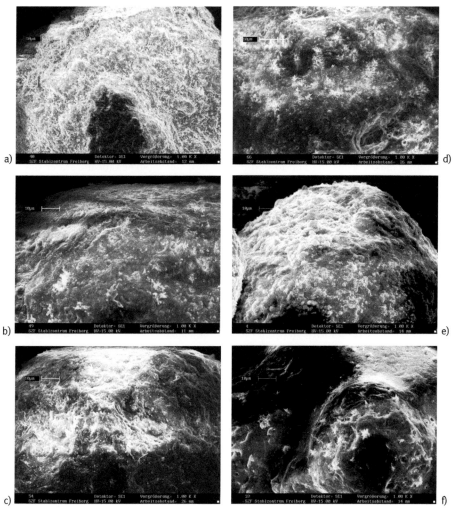

Figure 6.28: SEM images: grain surface, reclaimed mullite [6.29]
(new moulding material 1–5th reclaimed sand b–f)

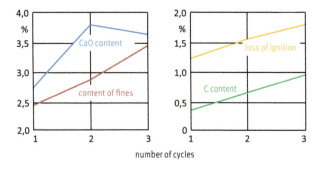

Figure 6.29: Variation of the content of fines, CaO and C content and the loss of ignition of cement molasses moulding material in several reclamation cycles [6.30]

components of the reclamation unit are the screw classifier and the actual cleaner with counter rotating impellers. These work on the basis of the mutual rubbing and cleaning of the sand grains through agitation of the rubber coated wheels. The author looks at the problems of the wet reclamation process including sludge removal, and recommends the allowance of water evaporation in the settling tanks for economic reasons.

Reuther makes it his mission to find Reclamation Technologies for the Water Glass Ester and Cement Form Process [6.32] and to practically test them. In the area of cement moulding material he examines three possible reclamation technologies: The fluidized bed cleaner, a vibration crusher and a pneumatically operated Jet Reclaimer. The system operating on a single stage vibration crusher yields no useable results while some success was achieved both with the small scale fluidized bed cleaner and the Jet Reclaimer. The form mixture used consists of 90 mass % silica sand, 10 mass % RNA 42.5 cement, as well as 5 mass % water, resulting in a theoretical dust content of 10 % (in essence, this is part of the cement portion). Figure 6.31 shows the dust losses and the remaining grout solids content in the reclaimed sand during reclamation periods from five to fifty minutes in the fluid bed cleaner. It is clear that the cement sand reclamation is possible, but long reclamation periods are necessary. Only after forty minutes of reclamation are strength values of over 80 % (of new sand values) achieved (figure 6.32). For practical use this means that new sand addition in the range of 30 % is required.

For applications with a low level of requirements, for example as backfill, vibration crushers can be worked with according to [6.32]. For more demanding applications of these materials this process is not appropriate. The Jet Reclaimer represents a practical solution in this area. Trial batches were reclaimed without predrying as well as with fifteen minutes predrying time at 150 °C. This was to evaluate the brittleness of the binder shells for improvement of reclamation behaviour. The achieved levels of recla-

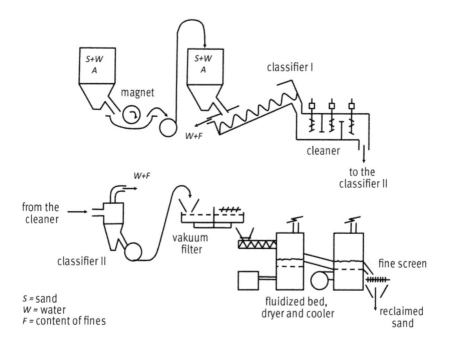

Figure 6.30: Wet reclamation system, schematic [6.31]

mation based on the electrical conductivity are shown in figure 6.33. Figure 6.34 shows the realized strengths with 100 % reclaimed sands. In summary, evaluations of the pneumatically operated Jet Reclaimer established the following [6.32]:
1. Predrying of used cement sands for the purpose of binder shell embrittlement is not necessary.
2. The required quality criteria can be met with 100 % used sands with a reclamation times of between forty and sixty minutes
3. In shorter cycle times it must be assumed that a 20 % to 30 % addition of new sand is required.

Summary of reclamation process

The difficult reclaimability of inorganically bonded used sands is often an argument as to why application of this occupational and environmentally friendly process is not in use today. The counter argument is then usually the much simpler reclaimability of or-

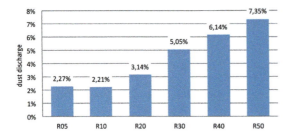

Figure 6.31: reclamation of used cement sands in the fluidized bed cleaner, dust discharge and content of fines in reclaimed sand (reclamation time at x – axis) [6.32]

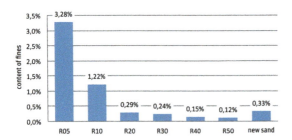

Figure 6:32: Relative compression strengths of reclaimed cement sand in comparison to new sand mixtures (reclamation time at x – axis) [6.32]

ganically bonded spent sands, for example, from the furan or phenolic resin process. Therefore, one must get an accurate perspective: Inorganically bonded used sand reclamation and the re-use of these sands in the moulding material and core production was already in practice decades ago as evidenced by a variety of sources quoted here. With the decline of these of process, the available equipment and know-how has been lost in favor of the more productive and superior technological properties of organic systems. At least in Germany, there is no reference system for water glass used sands. In other European countries, such as Czech Republic and Norway this is also the case. Furthermore, the supposedly much simpler reclamation of organically bonded

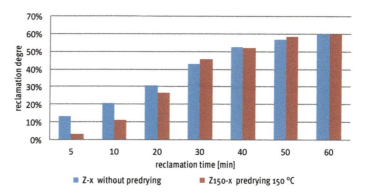

Figure 6.33: Reclamation degree cement moulding material, Jet Reclaimer [6.32]

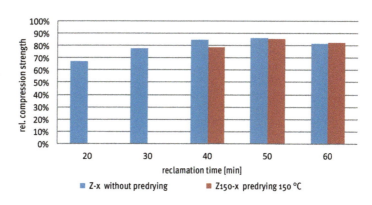

Figure 6:34: Relative compression strengths for new sand mixtures, Jet Reclaimer [6.32]

used sands is often a fallacy. One of the most persistent problems in large cast production using furan resin moulding materials is unacceptably high loss of ignition leading to increased incidence of gas bubbles or pinholes in the castings. This is offset by increased new sand addition. Looking at the used sand treatment process in detail, one will usually find a crushing unit such as a spiral crusher installed after the unpacking grate as well as a dust extractor. The very important dust extraction stage is often operated at low power in order to obtain as little dust as possible for monitored hazardous waste specifications. This reduces landfill costs but increases quality costs as well as scrapping expenses. Clearly said, this means that for proper reclamation, a reclamation device, from one of the options described at the beginning of this section, one that frees the moulding material grains from the spent binders, must be used.

There are a number of positive steps developed in the last twenty years in the field of inorganic used sand reclamation that are discussed in this section. Start up problems are unavoidable with any start-up technology, however the risk of failure is relatively low. One must only find the courage to begin ...

Literature – Chapter 6

[6.1] Weller, E., Möglichkeiten und Grenzen bei der Regenerierung von Gießereialtsanden GIESSEREI 76, 1989, Nr.10/11, S. 350–358
[6.2] Klat, F., Wiederverwendung von Kohlensäure-Erstarrungssanden, Slevarenstvi 10, 1962, Nr. 8, S. 306–308
[6.3] Palmer, B. V., Ingenieurs-Aspekte beim Kühlen, Transportieren, Konditionieren und Regenerieren von Formsanden, Foundry Trade Journal 1971, Nr. 2, S. 7–18
[6.4] Roberts, M., Die Anwendung des Ester-Selbsthärteverfahrens auf der Basis von Wasserglassanden, Regenerierungskosten, Meehanite-Report, Rapallo Juni 1972, S. 29–32
[6.5] Flemming, E., Literatursammlung zur Regenergierung wasserglasgebundener Altsande, Freiberg 1989
[6.6] Jennes, N., Möglichkeiten der Regenerierung wasserglasgebundener Gießereisande, Gießereitechnik 21 (1975), Nr. 12, S. 408–410
[6.7] Röhrig, G., Zur Wirtschaftlichkeit des CO_2-Verfahrens bei Verwendung von regeneriertem Kernsand, GIESSEREI 62 (1975), Nr. 5, S. 99–105
[6.8] Flemming, E., Schmidt, J., Wurzbacher, D., Erfahrungen und Wirtschaftlichkeitsbetrachtungen zum Einsatz einer Wirbelschichtanlage in einer mechanisch-pneumatischen Regenerierungsanlage, Gießereitechnik 27 (1981), Nr. 7, S. 213–216
[6.9] Vernay, P., Altsand-Rückgewinnung von Formstoffen für das Wasserglas-Ester-Verfahren, Hommes et Fonderie 1982, Nr. 124, S. 19–22
[6.10] Chahan, K., Cobett, T., Regenerierung selbsthärtender wasserglasgebundener Sande, Modern Casting 3 (1983), Nr. 3, S. 30–31
[6.11] Lissell, E. O., Regenerierung von wasserglasgebundenen Sanden, British Foundrymen 75 (1982), Nr. 2, S. 20–25
[6.12] Tilch, W., Gottschalk, J., Schulz, H., Heintze, H. G., Mechanische Regenerierung von Altformstoffen – Möglichkeiten und Grenzen, Gießereitechnik 32 (1986), Nr. 11, S. 331–335
[6.13] Polzin, H., Nitsch, U., Tilch, W., Flemming, E., Regenerierung anorganisch gebundener Altsande in einer mechanisch arbeitenden Pilotanlage, GIESSEREI-PRAXIS 1997, Nr. 23/24, S. 500–507
[6.14] Pohl, P., Beispiele zur Regenerierung von Gießereialtsanden, GIESSEREI 77 (1990), Nr. 21, S. 659–663
[6.15] Polzin, H., Flemming, E., Nitsch, U., Tilch, W., Jelinek, P., Miksowsky, F., Grundlagen und praktische Erfahrungen zur Regenerierung von Altsanden aus dem Wasserglasverfahren, GIESSEREI 85 (1998), Nr. 11, S. 53–59
[6.16] Döpp, R., Alekassir, A., Xiao, B., Beitrag zum mechanischen Regenerieren kaltharz- und wasserglasgebundener Formstoffe, GIESSEREI 82 (1995), Nr. 4, S. 111–115
[6.17] Spillner, A., Erfolgreiches Modellprojekt zur Regenerierung von Wasserglas-Altsand in Aluminiumgießereien, GIESSEREI-PRAXIS 1998, Nr. 9, S. 366

[6.18] Polzin, H., Nitsch, U., Flemming, E., Tilch, W., Regenerierung von Wasserglasaltsanden in mehreren Umläufen, GIESSEREI-PRAXIS 1999, Nr. 3, S. 101–105
[6.19] Polzin, H., Tilch, W., Praktische Erfahrungen zur mechanischen Regenerierung wasserglasgebundener Altsande, GIESSEREI-PRAXIS 2004, Nr. 5, S. 191–197
[6.20] Müller-Späth, J., Westhoff, E., Schädlich-Stubenrauch, J., Sahm, P. R., Öko-Sandregenerierung, GIESSEREI-PRAXIS 1994, Nr. 11/12, S. 308–316
[6.21] Polzin, H., Tilch, W., Herstellung von Kupferguss- und Gusseisen-Bauteilen mit anorganischen Formstoffen, GIESSEREI 2006, Nr. 10, S. 44–51
[6.22] Polzin, H., Tilch, W., Neue Erkenntnisse zum Umlaufverhalten regenerierter Wasserglasformstoffe, GIESSEREI-PRAXIS 2004, Nr. 11, S. 415–420
[6.23] Fan, Z. T., Huang, N. Y., Dong, X.P., In-house reuse and reclamation of used foundry sands with sodium silicate binder, Int. J. of Cast Metal Research, 17, 2004, Nr. 1, S. 51–55
[6.24] Fan, Z., Huang, N., Wang, H, Dong, X., Dry reusing and wet reclaiming of used sodium silicate sand, CHINA FOUNDRY 2, 2004, Nr. 1, S. 38–43
[6.25] Genzler, Ch., Anorganische Formstoffbindemittel – Silikate, GIESSEREI-PRAXIS 2005, Nr. 3, S. 8994
[6.26] Danko, R.S., Experiences gathered during reclamation of used water glass and bentonite sands in extra low and ambient temperatures, Int. J. of Cast Metals Research 23, 2010, Nr. 2, S. 92–96
[6.27] Polzin, H., Jaruszewski, P., Müller-Späth, J., Die Regenerierung von Wasserglas-Ester-Altsanden in einer mittelständischen Aluminiumgießerei, GIESSEREI 99 (2012), Nr. 4, S. 70–81
[6.28] Pavlovsky; J., Herecova, L., Micek, D., Mucha, M., Knurova, L., Praus, P., Vaskova, I., Polzin, H., Die Bestimmung von Natriumazetat in Wasserglas-Ester-Formstoffen mit Hilfe der kapillaren Isotachophorese, GIESSEREI-PRAXIS 2012, Nr. 3, S. 62–67
[6.29] Nicklisch, M., Die Regenerierung von wasserglasgebundenen synthetischen Hohlkugelmulliten, Studienarbeit, TU Bergakademie Freiberg, 2011
[6.30] Schmidt, J., Formstoffregenerierung – Möglichkeiten und Probleme, Gießereitechnik 24 (1978), Nr. 3, S. 87–90
[6.31] Zimnawoda, H., Verfahren zur Sandrückgewinnung, GIESSEREI 59 (1972), Nr. 20, S. 594
[6.32] Reuther, R., Vorbereitung des Einsatzes einer Anlage zur Formsandregenerierung im MMG für die eingesetzten anorganischen Formstoffsysteme, Diplomarbeit, TU Bergakademie Freiberg 2003

7 The Influence of Inorganic Binder on Clay Bonded Circulation Moulding Materials

One of the biggest advantages of bentonite bonded moulding materials is the almost complete recoverability of incurred used sands, their recyclability and their re-use for making moulding materials. The concept of 'bentonite circulation moulding material' describes this fact clearly. The binder system of bentonite water loses its binding ability due to loss of swelling capacity only in the immediate vicinity of liquid metal. The so-called 'clay dead burning' (Oolithe content and dustiness) sets, depending on bentonite quality, in the temperature range between 500 °C and 600 °C. The absolute amount of dead burned bentonite can be minimized by targeted control of the system, via the change in the ratio of the moulding material to metal or the dwell time of the casting in the moulding material leading up to unpacking. Another factor in the deterioration of the circulation characteristics of bentonite bonded moulding material sand is the supply of used core sands after the separation of the casting and the moulding material. Due to the good decay behaviour of some core moulding materials, advanced unpacking technologies and the high productivity of foundry process, a 100 % separation of both moulding material cycles is impossible. *Tilch* states in [7.1] that 'changes in current behaviour of the clay bonded moulding material sand are caused by thermal, mechanical and chemical influences'. The influence on the binding properties of bentonite can therefore, result from core gases of pyrolysis products of core binders, carbonization gases of moulding material additives (primarily coal dust) or the inflowing old core sand itself. The effects of these processes are according to *Hofmann* [7.2], the desactivation, acidification and salinisation of alkaline bentonite. Or, as *Boenisch* describes [7.3], a bentonite crusting (the so-called 'cover' effect) when using organic core binder systems. Since the inorganic core binder systems discussed here are sodium silicate based, the mechanism of injury ('over activation') of the bentonite is made clear by the activation curve (figure 7.1). To assess the effect of cover effect, *Boenisch* developed the so-called 'generator receptor process' [7.4], which will be discussed at later point in terms of the behaviour of inorganic material. In the experiment shown in figure 7.2, the generator moulding material is heated hot gas burners at 1000 °C. The resulting combustion products then pass into the receptor moulding material above and can condense there. By processing the receptor moulding materials and determining the technological properties such as the strength, the influence of the pyrolysis gases can be established. Characteristic size of the process is the generator/receptor ratio (GRR), which describes the ratio of the moulding material quantities used: A GRV

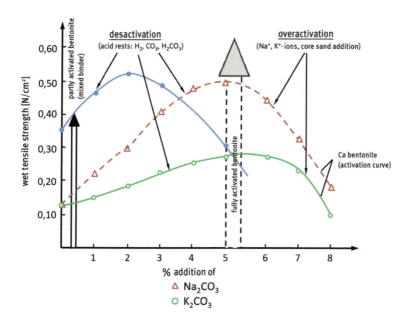

Figure 7.1: Activation curve of bentonites – desactivation (or overactivation) by influx of ions (from core binders)

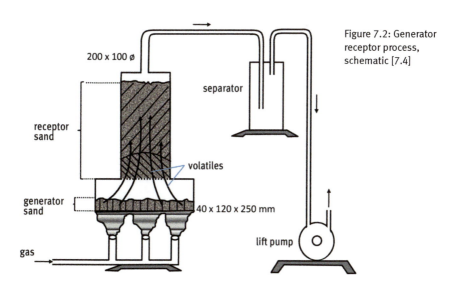

Figure 7.2: Generator receptor process, schematic [7.4]

The Influence on Clay Bonded Sands | 203

of 0.4 means that 40 g of core moulding material will pass its pyrolysis products to 100 g of bentonite bonded moulding material.

Pohl and *Sagmeister* in [7.5] add various used core sands for bentonite moulding material and investigate the properties. Table 7.1 shows the variation of various characteristics with the addition of increasing amounts of used core sands starting at 20 %. In figure 7.3 the effect of core residue is shown in the wet tensile strength. It increases with increases in used core sand but the damage caused by the water glass used sand falls lower. *Flemming* and *Tilch* investigate in [7.6], inter alia, the damage mechanisms of various core moulding material systems on the bentonite moulding material. They note that the carbonization gases from water glass bonded moulding materials cause relatively little damage. The entrance of property deterioration is therefore, attributable to the activation of the bentonite by NaOH. The wet tensile strength of bentonite bonded moulding material sand is the most sensitive to deactivation process. The car-

Table 7.1: Change in the properties of clay bonded moulding material sand (8 % bentonite, 4 % coal dust) with increasing additions of water glass CO_2 used core sands [7.5]

		Silica sand H32 8 % bentonite 4 % coal dust without core residue	Addition to water glass bonded core residues		
			20 %	40 %	60 %
Moisture	%	2,7	2,7	2,7	2,7
Mouldability	%	75	80	73	80
Density	g/cm³	1,53	1,53	1,53	1,52
Gas permeability		135	137	143	137
Compression strength	N/cm²	14,32	12,16	9,32	8,24
Splitting strength	N/cm²	2,69	2,16	1,67	1,27
Binding capacity	%	18,8	17,8	17,9	15,4
Flowability	%	45	45	49	49
Shatter-Index	%	71	68	58	56
Stripping parameter		4,9	5,5	6,1	6,7
Wet tensile strength	N/cm²	0,32	0,16	0,08	0,06
Compressive stress	N/cm²	28,44	26,48	22,55	18,63
Tilt error*)		89	166	282	311

*) Value < 150 – no surface defects

bonization from the water glass process therefore, has no significant influence on the wet tensile strength. *Brümmer* and *Lautzus* [7.7] also carry out investigations on the effect of core old sands and come to the conclusion that up to 12 % used core sands can accrue independent from the system without causing problems. While they surprisingly find property improvements at 15 % used water glass sand, the properties deteriorate at higher core sand additions (figure 7.4). Earlier however, *Granitzki* had found in [7.8] that with the inflow of used water glass sand strength increases of up to 5 % can be achieved, while at further increased amounts of water glass sand, strength reductions occur (figure 7.5). Wet tensile strength generally decreases with the influx of sodium silicate used sand A reduction of 70 % is caused by a 40 % addition! A way to counteract the overactivation is proposed here by the addition of nonactivated calcium bentonite.

The influence of used cores sands on green sand systems has been studied thoroughly in more mo-

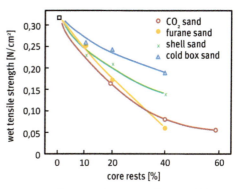

Figure 7.3: Change in the wet tensile strength from the inflow of various used core sands [7.5]

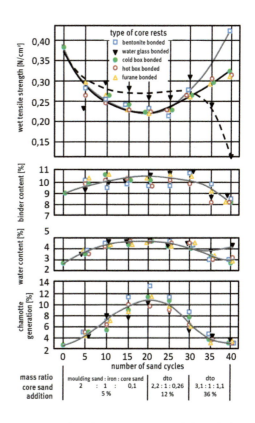

Figure 7.4:
Effect of different proportions of core sand supply on the properties of a clay bonded moulding material [7.7]

The Influence on Clay Bonded Sands | 205

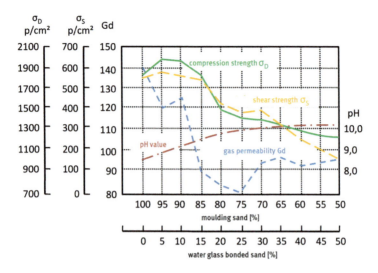

Figure 7.5: Change in the moulding material properties by the addition of water glass used sand [7.8]

dern times by authors such as *Buse* [7.9]. She starts with a mould mixture of 8 % Greek and Bavarian bentonite prepared to a compressibility of 45 +/- 2 %. In addition to PUR cold box, resol ester and resol CO_2 core sands, two water glass CO_2 used sands (3 % binder content) are added. The amount of core sand is between 10, 25 % and 50 %. The influence of pyrolysis products from the used core sand on the clay bonded material (generator receptor process) was then examined, as well as the impact on the mould properties from direct addition of used sand. In assessing the technological properties, it was found that the incoming damage as a function of added used core sands varies greatly. It confirmed that the (permissible) inflow quantities for the individual used sands approximately correlated to the existing literature on the subject. A core sand inflow of 50 % was determined to be inadmissible for all investigated core moulding materials. The results for the water glass CO_2 used core sands when using Greek and Bavarian bentonite are shown in table 7.2. By the evaluation of the generator/receptor tests with a GRV of 0.6, lower values were achieved here than shown in existing literature with some results deviating greatly.

Flemming prefers the procedure just described [7.10], but only uses water glass CO_2 sand used sand for his investigations. Three different water glass binders were included, two modified binder systems and an unmodified water glass with a modulus of 2.5. In order to simulate the various possible 'thermal history' of the core moulding materi-

Table 7.2: Changes in properties of bentonite-bonded moulding materials through the inflow of different amounts of water glass CO_2 used core sands [7.9]

sodium water glass CO_2 core sand		moulding material with greek bentonite (%)								moulding material with bavarian bentonite (8 %)							
directly		directly inflow				generator receptor process				directly inflow				generator receptor process			
		quantity of core old sand in the sand				resin : binder ratio				quantity of core old sand in the sand				resin : binder ratio			
directly	unity	0	10	25	50	0	0,2	0,4	0,6	0	10	25	50	0	0,2	0,4	0,6
wet tensile	[N/cm²]	0,397	0,456	0,320	0,122	0,416	0,445	0,456	0,435	0,525	0,484	0,395	0,166	0,503	0,496	0,527	0,493
green tensile	[N/cm²]	3,16	3,00	2,89	2,81	3,40	2,80	2,99	3,04	3,31	3,10	3,06	2,84	3,15	3,19	3,09	3,01
green compression	[N/cm²]	15,8	15,8	15,1	13,3	17,0	14,5	15,6	15,7	15,4	16,1	17,4	14,5	15,7	15,9	15,7	15,9
permeability		223	229	189	164	240	287	235	263	210	197	170	163	206	205	219	203
plasticity	[%]	78,6	80,9	78,1	79,2	78,9	79,1	77,6	79,9	74,8	72,7	72,3	70,7	78,9	75,3	73,8	74,6
compactibility	[mm]	0,37	0,41	0,32	0,23	0,37	0,43	0,41	0,42	0,38	0,37	0,31	0,24	0,35	0,37	0,38	0,37
shear	[N/cm²]	4,50	4,58	4,17	4,42	4,82	4,08	4,40	4,51	5,26	4,93	5,28	4,74	5,02	4,40	4,75	4,53
water content	[%]	2,20	2,51	2,85	3,13	2,35	2,41	2,29	2,37	2,18	2,33	2,36	2,74	2,21	2,25	2,34	2,39
pH value		10,39	10,74	10,87	10,94	10,40	10,14	10,04	10,04	10,26	10,44	10,62	10,80	10,30	10,25	10,27	10,13
LOI	[%]	0,76	0,90	1,10	0,90	0,80	0,63	0,61	0,59	0,65	0,73	0,69	0,77	0,74	0,72	0,75	0,75
content of fines	[%]	8,12	8,56	8,70	8,50	8,32	8,12	8,38	8,18	7,76	8,06	8,08	8,10	7,68	8,06	8,00	8,32
electr. conductivity	µS/cm	559	686	721	937	558	541	572	530	368	460	610	791	368	390	391	393

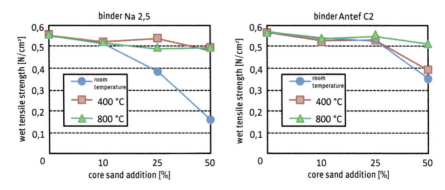

Figure 7.6: Wet tensile strength through inflow of water glass CO_2 used core sands, left-unmodified, right- modified binder [7.10]

als, thermally unstressed core sands (broken cores) and used sand that has been thermally loaded at 400 °C and 800 °C was used. Again, the direct influence of the core sand inflow to the green sand, as well as the effect of pyrolysis gases was evaluated. Figure 7.6 shows the change in wet tensile strength by inflow of 10, 25 % and 50 % used core sands. It is interesting to note here that the modified water glass addition causes no damage at 25 % inflow, while the unmodified water glass causes a significant reduction of wet tensile strength at 25 % addition. Accordingly, it is to be noted that the modified water glass binder is a system containing organic additives of approximately 20 %. The overall results of these investigations are given in table 7.3. Table 7.4 clearly shows that with the water glass CO_2 process that is, in the broadest sense of inorganic bonded moulding material process – it can be expected that no significant damage to the clay bonded moulding material sand occurs from the lack of pyrolysis gases, and in some cases, even slight improvements in properties are observed.

To sum up the problems of the direct influence of clay bonded moulding material systems through inorganic core old sands, it can be determined that inflow quantities of 10 % are generally harmless, while larger quantities have no impact only under certain conditions. Obviously, this is applicable only for the classic water glass CO_2 process. For newer inorganic binder systems, a similar behaviour is to be expected, but must be tested for each specific case. The risk is lower than is sometimes assumed. Also any negative effects can be delimited by targeted influencing of the core sand inlet such as core sand separation.

Table 7.3: Evaluation of different core sand inflow [7.10] (M1 and C2-modified Na 2.5 unmodified water glass)

		wet tensile	green tensile	shear	permeability	gas evolution	sintering	pH	AGS	content of fines	surface
M 1 / RT /	10 %	-	0	-	-	+	-	0	0	<	<
	25 %	--	--	-	--	+	--	-	>	<	<
	50 %	--	--	--	--	+	--	-	>	<	<
M 1 / 400 °C /	10 %	0	0	0	-	+	-	-	>	<	<
	25 %	0	-	+	0	+	--	-	>	<	<
	50 %	--	-	0	-	+	--	-	>	<	<
M 1 / 800 °C /	10 %	-	0	+	-	+	-	0	>	<	<
	25 %	-	0	0	0	+	-	-	>	<	<
	50 %	-	--	--	+	+	--	-	>	<	<
C2 / RT /	10 %	0	0	+	0	+	0	+	0	<	<
	25 %	0	-	0	0	+	-	-	>	<	<
	50 %	--	--	--	--	+	--	-	>	<	<
C2 / 400 °C /	10 %	0	0	0	-	+	-	0	0	<	<
	25 %	0	0	-	0	+	-	-	>	<	<
	50 %	--	-	-	--	+	--	-	>	<	<
C2 / 800 °C /	10 %	0	0	0	-	+	0	-	0	<	<
	25 %	0	-	0	0	+	-	-	>	<	<
	50 %	-	-	-	0	+	--	-	>	<	<
Na 2,5 / RT /	10 %	-	0	0	-	+	-	-	0	<	0
	25 %	--	--	0	--	+	-	-	0	<	0
	50 %	--	--	--	-	+	--	-	0	<	0
Na / 400 °C /	10 %	-	0	+	-	+	-	-	0	<	0
	25 %	0	-	-	-	+	-	-	0	<	0
	50 %	-	-	-	-	+	--	-	0	<	0
Na / 800 °C /	10 %	-	0	0	0	+	0	-	0	<	0
	25 %	-	0	0	0	+	-	-	0	<	0
	50 %	-	--	--	0	+	-	-	0	<	0

+   improvement in comparison to bentonite bonded sand
0   no change
-    small decrease (max. 10 %) to the bentonite bonded sand
--   higher decrease (more than 10 %) to the bentonite bonded sand
>   increased value
<   decreased value

Table 7.4: Results of the generator receptor process [7:10]

Test size	Clay bonded sand	Gassed with Antef M1	Gassed with Antef C2	Gassed with Na 2,5
σ_{NB} [N/cm²]	0,555	0,510	0,502	0,520
σ_{ZB} [N/cm²]	334	331	335	332
σ_S [N/cm²]	4,35	4,05	4,40	4,70
Gd [Skt]	212	205	208	215
Ge [ml]	7,2	5,6	6,9	3,7

σ_{NB} wet tensile strength
σ_{ZB} green tensile strength
σ_S green shear strength
Gd gas permeability
Ge gas evolution

Literature – Chapter 7

[7.1] Tilch, W., Das Umlaufverhalten tongebundener Formstoffe, Gießereitechnik 33 (1987), Nr. 12, S. 363–368
[7.2] Hofmann, F., Die thermische Beständigkeit von Bentoniten und ihre Bedeutung für das betriebliche Verhalten bentonitgebundener Formsande, GIESSEREI 50 (1963), Nr. 5, S. 123–131
[7.3] Boenisch, D., Über den Einfluß von Cold-Box-, Hot box- und Croningkernen auf die Eigenschaften bentonitgebundener Formsande, GIESSEREI 64 (1977), Nr. 21, S. 549–554
[7.4] Boenisch, D., Die Kondensation von Harzdestillaten in der Sandform, GIESSEREI 64 (1977), Nr. 8, S. 207–212
[7.5] Pohl, W., Sagmeister, H.D., Veränderung der Formsandeigenschaften durch den Kreislauf und durch Zufluß von Kernsanden verschiedener Art, GIESSEREI-PRAXIS 1974, Nr. 5, S. 83–96
[7.6] Flemming, E., Tilch, W., Beitrag zur Erhöhung der Gussstückqualität und einer verbesserten materialökonomischen Verwendung von Gießereiformstoffen; Dissertation B, TU Bergakademie Freiberg, 1981
[7.7] Brümmer, E., Lautzus, W., Einfluß von Kernsanden auf den bentongebundenen Umlaufsand, GIESSEREI 64 (1977) Nr. 23, S. 612–616
[7.8] Granitzki, K.-E., GIESSEREI 17 (1965), Nr. 1, S. 1–16
[7.9] Buse, K., Untersuchungen zur Einflussnahme von Kernsanden auf die Eigenschaften bentonitgebundener Formstoffe, Diplomarbeit, TU Bergakademie Freiberg, 1994
[7.10] Flemming, E., Polzin, H., Untersuchungen zum Einfluss von wasserglasgebundenen Kernaltsanden auf tongebundene Formstoffsysteme, GIESSEREI 84 (1997), Nr. 16, S. 19–23

Index

A

abrasion resistance *77, 99, 121*
abrasive drums *166*
accelerator *35, 78f, 83ff*
acetic acid *10, 89, 94, 102, 109*
acid *16, 21, 32, 49, 94, 100, 171*
acidification *124, 202*
activation curve *202, 203*
additives *5, 25ff, 35, 37, 47, 52, 59, 78f, 83, 97ff, 107, 116, 122f, 130ff, 146, 163f, 179, 202, 208*
adhesion *17, 48, 92, 136, 171*
adsorption layer *21*
agglomeration processes *31*
aggregation *50, 101*
AlO_4 *108*
air acceleration *167*
alcohol coating *58, 119*
alcoholic binders (ethyl silicate) *13, 137*
alkali anions *20*
alkali carbonate *15*
alkali content *15, 32*
alkali ions *18*
alkali metal oxide *16*
alkali metal silicate *16, 30*
alkali metal silicate solutions *25*
alkali oxide *146*
alkali silicate *14ff, 31, 38, 58, 168*
alkali silicate solution *9, 14ff, 20, 29, 43, 62*
alkaline additives *21*
alkaline component *44*
alkaline geopolymer *108*
alkaline oxide *14*
alkaline salts *14*
alkalinity *31, 44*
alloyed cast iron *182*

AlO$_4$ tetrahedra *36, 108*
alumina *32, 79, 150ff*
alumina cement *32, 34*
aluminium bronze propellers *12*
alumino silicate sand *147, 191*
aluminosilicate solution *36*
aluminium *11, 12, 36, 65, 105, 107ff, 117ff, 131, 135, 159, 161, 188, 192*
aluminium bronze alloy *178*
aluminium casting *12, 38, 44, 53, 58, 108, 110, 120, 187, 189*
aluminium foundries *11, 105, 164, 174*
amorphous silica *31, 48*
anhydrous systems *53*
aqueous silica *13, 43, 137*
aqueous systems *13*
atomic absorption *173*
attrition *168f, 185*
autoclave *15f, 139, 141*
automotive castings *53*
auxiliary materials *146*
AWB® process *41, 131*

B

backfill material *79, 11*
backfill sand *47*
ball mills *166*
Baume *14*
BeachBox® process *135*
bentonite *5, 10, 62, 73, 163, 169, 172, 174, 177, 186, 202ff*
bentonite crusting *202*
binder bridge *17, 50, 54, 69, 74, 95, 98, 114, 120, 125, 130, 186*
binder film *20, 22, 52, 61f, 98, 124, 171*
binder system *5, 9, 11ff, 26, 28, 30ff, 36ff, 41ff, 60, 70f, 79, 88f, 94, 99, 103, 105, 107ff, 114ff, 125ff, 132, 135ff, 142, 148, 151, 162, 168f, 202, 206, 208*
binder to hardener ratio *94*
blast furnace cement *32, 33*
blast furnace slag *33, 34*
borate *38, 117, 136*
boric acid *47*

boron *102*
box-less operation *73*
brittleness *52, 186, 196*
burning *140f, 147, 202*

C
CaCl$_2$ *78, 84ff*
calcination *39*
calcium *34, 83, 123*
calcium hydroxide *73*
calcium oxide *32*
calcium silicate hydrate *73*
capillary pores *74ff*
carbon dioxide *23, 28, 30, 43ff, 52ff, 58, 60ff, 91, 107f, 114f, 121, 124, 128f, 150*
carbon dioxide gassing *11, 59, 62f, 44, 56, 114, 124ff, 129*
carbon dioxide-air mixture *44*
carbonation solidification *43*
carbonic acid (H$_2$CO$_3$) *49, 54, 100, 102*
carbonization gases *202, 204*
cascade scrubber *168*
CaSO$_4$*2H$_2$O *39*
cast iron *51, 65, 73, 89, 91, 105, 135, 146f, 176, 182*
cast steel *9, 47, 90, 108*
casting surface *70, 97, 149*
catalyst *32*
cement *5, 9, 12, 32ff, 43, 72ff, 124, 193, 196ff*
cement gel *74f*
cement molasses moulding material *196*
cement molasses moulding method *194*
cement moulding *9, 72, 75, 79, 80, 172, 196*
cement moulding process *12, 24, 33ff, 41, 71ff, 83, 86ff*
cement paste *73f*
cement stone *74f*
chemical binder *32*
chemical bonding *23f*
chemical curing *10, 23, 43, 118*
chemical curing system *9*
chemical modification *25, 26*

chemical reaction   50, 55, 61, 73, 115, 157
chemically cured moulding material system   41
chemically curing binder systems   5, 30
chemically curing moulding method   41
chemically hardened moulds and cores   43
chromite   28, 147ff, 157, 159ff, 171, 173
chromium oxide   122
clay   5, 9, 47, 89, 125
clay bonded circulation moulding material   202
clay bonded moulding material   69, 202, 204ff
clay dead burning   202
cleaning requirements   46
CO_2 fumigation   27
CO_2 gassing   49, 56, 61, 127
CO_2 process   5, 9, 11, 41, 168, 177, 183, 208
coagulation   21f, 31, 48ff, 54, 91ff, 102, 127
coagulation process   50, 102
coagulation threshold   28f, 92
coal dust   47, 79, 122, 202, 204
coating   58, 97, 119, 132, 137, 139f
cohesion   17
cohesive strength   29
cold air   38, 56, 65
cold box gasification method   43
cold box process   5, 9, 54
cold self-curing process   9, 69, 89
cold self-curing   24, 69, 71, 88
cold self-curing moulding process   10, 71
cold self-hardening method   71f, 96, 108
collabsibility   28, 92, 102
colloid   17, 21, 23, 50, 63, 92, 141
colloid ions   5, 20
colloid solution   31, 49, 54, 92
colloidal silica   31, 94
colloidal solution   31, 49, 54, 92
compaction   5, 10, 76, 90
component balance   164
composite cement   33

compressive strength *30, 33f, 45, 51, 57, 59, 84, 86f, 126*
concentration *17ff, 31, 62, 114, 191*
concrete *36, 73, 76f, 83, 132, 169*
concrete plasticizers *83*
condensation *21, 25, 27f, 31, 50, 62f, 102, 119, 121, 130, 137*
condensation reaction *31, 115, 119*
conductivity *29, 142, 147, 149, 159, 176, 184f, 188, 190, 192f, 197, 207*
contact angle *29*
continuously working furnaces *15*
cooling *51f, 59, 98, 116, 118, 137, 142, 149f, 164, 166f*
cooling rate *50f*
copper foundries *11*
Cordis® *42, 115, 117ff*
Cordis® process *41*
core *9ff, 23, 26, 28, 32, 37f, 43ff, 50, 53, 55f, 58ff, 65, 69ff, 74, 78f, 89f, 98f, 107, 110f, 114ff, 149, 163f, 166, 169, 174, 183, 185, 202ff*
core box *10, 24f, 38, 46, 56, 58, 63, 70, 114f, 117f, 120, 130, 132, 135, 153, 161*
core box vents *120*
core making *41, 164*
core production *5f, 9, 11, 26, 37, 46, 58f, 63, 73, 76, 91, 97, 99, 107, 110, 114f, 117, 122, 126f, 132, 147f, 168, 175, 198*
core shooting machine *37, 45, 117, 130ff, 135, 160*
corrosion *38*
corundum *147, 151, 160f*
cracking *45, 53, 99, 140*
critical humidity *56*
crystal water *75, 98f*
cupola furnace slag *122*
curing *9f, 20ff, 24, 34f, 41, 43ff, 47f, 52f, 56, 61ff, 70, 72ff, 77, 83, 86, 89ff, 95, 97, 99ff, 108f, 115, 118, 120ff, 126, 129ff, 157, 162f, 171*
curing accelerator *78, 87*
curing time *9, 34, 72, 77, 97, 101f, 118, 120, 122f, 126, 186*
cycle time *5, 38, 115f, 121, 130f, 135, 197*
cylinder block *63, 124*
cylinder head *53, 115, 117f, 120*
cylinder head core *137*

D

deactivation 204
dendrite arm spacing 116
density 14f, 17f, 20, 48f, 91, 127, 136, 138, 146f, 152, 161, 193, 204
density change 17, 146
dextrin 26, 79
diacetin 90, 95, 97, 102f, 171, 192
dicalcium silicate 35
dielectric heating 121
diethylene glycerol diacetate 95
dilatometer testing 62
dimensional stability 24, 77, 122, 159
dipole character 53
disilicic acid ($H_2Si_2O_5$) 54
disposal procedure 11
drill core 11
dry sand process 53
drying 9, 13, 19, 21, 23f, 27, 30f, 41, 44ff, 49f, 52, 55, 58, 61ff, 97ff, 114, 212ff, 135, 137, 139ff, 157
drying chambers 63
drying times 46, 62, 126, 129, 141
DTA/DTG 63
DTG gradients 63
dustiness 202
dynamic viscosity 29

E

electrical conductivity 188, 190, 192f, 197
electrokinetic potential difference 21
electrokinetic potential (zeta potential) 21, 29, 49, 54, 92, 94
electrolyte 21, 49, 92, 189, 191
endoscopic examinations 121
endothermic peaks 63
environment conditions 12
environmental behaviour 77
erosion 77
ester hardener 95, 99, 101, 171
ester hardening 27, 121

ester number *100ff*
ethyl silicate *13, 43, 137*
ethylene glycerol diacetate *95*
ethylene glycol *95*
ethylene glycol diacetate *90*
expansion errors *146*
ferrosilicon (Nisyama method) *9, 28, 89*

F
final strength *69f, 80, 96, 105, 125, 171*
fire resistance *31, 79, 193*
fireclay *43, 136, 150ff, 159ff*
flowability *25f, 53, 60f, 99, 107, 118, 131, 204*
flow mixer *71, 78*
fluid bed cleaners *66*
fluidity *18, 60, 90, 117*
fluidized bed cleaner *172, 178, 180, 191, 196, 198*
fluidized bed furnaces *168*
free energy *49*
free water *52, 62, 74*
free water content *29*
fumigation *27, 46, 54f, 114f, 118, 133*
furan resin process *171*
furan resins *148*

G
gas bubbles *88, 117, 166, 199*
gas permeability *73, 76, 79ff, 84f, 87, 126f, 138, 166, 24, 210*
gas-curing moulding method *41*
gaseous hardener *41*
gasification *41, 43, 52, 110*
gasification process *43ff*
gassing *2111, 30, 41, 44, 46ff, 52ff, 61ff, 65, 114, 120, 124ff, 149f, 155, 177*
gel films *52*
gel structure *20, 54*
general engineering *76*
generation of gases *26*
generator receptor process *202f, 206f, 210*

geopolymer *236, 40, 108f, 110*
geopolymer binder *36, 41, 108, 110*
geopolymer process *108, 110*
geopolymers *35, 71*
Gisacodur *10, 90, 112*
glycerol *95, 101, 105f*
glycerolacetat *102*
grain sizes *54, 118, 138*
granulometric properties *146*
gray iron *90*
green sand moulding process *69*
gypsum *10, 12, 39*

H

hardening *9, 25ff, 31, 38, 41, 43, 46, 48ff, 55, 60, 62ff, 72f, 76ff, 83ff, 92, 94, 100f, 103, 109f, 114, 118, 120f, 123, 125ff, 130f, 149, 153, 166*
hardening accelerators *83ff*
hardening process *48, 99*
HCl *92*
heated core *37, 56, 69, 115, 117*
heated water vapor *65*
high frequency energy (microwave drying) *62*
high-alloy steel castings *76*
high-temperature behaviour *70, 129*
high-temperature properties *50*
hot air *30, 37, 41, 56, 62f, 65, 68, 115, 118, 120, 123, 175*
hot box method *24, 37, 49*
hot box process *114, 149, 151, 157ff, 161*
hot curing moulding process *14*
hot curing process *114*
hot distortion *149, 157ff*
hot formability *157*
hot or warm tools *62*
humid environment *56, 114*
hydrated silicon dioxide *43f*
hydration *32, 35, 73ff, 80f, 83ff*
hydraulic module *32*
Hydrobond® process *132*

hydrogen bonds *62*
hydrolysis *31f, 48, 94f, 100*
hydrolysis reaction *21, 31*
hydrothermal method *15f*
hydrothermal process *16*
hydroxides *35*
hygroscopic *56*
hygroscopicity *123, 127*

I

impurities *15, 18, 146*
initial sintering temperatures *124, 172, 174, 177, 189*
initial strength *33f, 54ff, 58, 80, 120, 123, 129*
inorganic additives *26, 97, 99, 143*
inorganic binder systems *5f, 9, 11ff, 41, 56, 71, 89, 11 , 114f, 135, 137, 168, 208*
inorganic hot box method *37, 41*
inorganic moulding processes *41*
inorganic processes *11*
Inotec® *42, 115ff*
Inotec® process *41, 115*
internal self-curing *41*
investment casting *9, 13, 31, 41, 134, 136ff*
investment casting shell moulding materials *13*
iron *11, 47, 59, 90, 105, 107, 110, 185*
iron castings *38, 45, 58, 103, 117, 164, 172, 188f*
iron foundries *45, 176*
iron oxide *32, 122*
isoelectric point *21, 50, 56*
Isotachophoresis *189, 192*

L

lamellar graphite *91*
land fill *77*
lean concrete *76*
length change *146*
light metal casting *38*
limonite *50*
liquid ester hardener *23, 28, 61*

liquid hardener 69, 90
lithium (Li) 16, 26, 123f
lithium silicates 16
loam 9
lost wax 9, 137, 139
lost wax casting process 13
lost wax casting method 43
lump crushing 164

M

machine tool castings 76
magnesium 99, 123, 148
magnesium hydroxide 73, 99
magnesium sulfate ($MgSO_4 \cdot H_2O$) 37f, 40, 135f
magnetic field treatment 28f, 97f
magnetic influence 28
magnetic treatment 29, 97
malleable iron 47
Me_2O 14
mechanical properties 11, 52, 142
mechanical reclamation 166f, 172, 178f, 183, 187f
melting process 16
metal separation 164
metal to sand ratio 70
Mg alloy 47, 188
$Mg(OH)_2$ 97, 99
MgO 97, 131, 150
$MgSO_4$ 37f
micelle 18, 20f, 49, 54, 92
microwave curing 38, 123
microwave drying 24, 62, 121 ff, 135
microwave drying method 121
mineralization 146, 149
modification changes 146
modulus 14 ff, 23f, 26f, 29, 32, 44f, 48ff, 52ff, 56, 59, 62f, 91, 96f, 102f, 121, 123ff, 171, 206
moisture absorption 53, 124
molar ratios 14, 63

molasse *26, 59, 78f, 172, 194, 196*
molybdate reactive curves *27*
mono silicate *22f*
mono system *182*
monoacetin *94f*
monomeric *20*
mould box *46, 70, 105*
mould production *31, 46, 97, 105, 134, 136*
moulding material *5, 10ff, 17f, 22ff, 30, 36f, 39, 41, 44f, 47f, 50ff, 54ff, 61, 66, 69ff, 73, 75ff, 89ff, 94, 97ff, 102f, 108, 110f, 114f, 117, 121, 124f, 127, 130ff, 146ff, 157ff, 166ff, 171ff, 177ff, 183, 186ff, 191ff, 198f, 202ff*
moulding mixtures
mullite

N

Na_2O *13, 17f, 25, 27, 31, 163, 171ff, 177f, 183ff, 192ff*
$Na_3P_4O_9$ *27*
Na_3PO_4 *27f*
$Na_4P_4O_{12}*4H_2O$ *27*
$Na_5P_3O_{10}*6H_2O$ *27*
natural sand *73*
neutralization *44, 54*
Nishijama method *89*
NMR spectrum *22f*
non-flammable *53*
non-offensive in odor *53*
non-toxic *53*
non-ferrous casting *10, 132*
normal strength *34*

O

oil bonded moulding material *171*
oil sand process *53*
olivine sand *90, 147f, 151ff, 160, 173*
oolithe content *202*
open curing *70*
organic additives *26, 37, 59, 97, 99, 143, 208*
organic binder systems *5ff, 9ff, 31, 41, 53, 56, 70f, 79, 88f, 94, 103, 110,*

114f, 119f, 135, 137, 168, 208
organic components   10, 26, 59f
organic condensates   116
organic system   5, 53f, 70f, 119f, 132, 198
over activation   202
over-gassing   52
oxides   35, 43, 146, 159

P

particle   20f
particle size distribution   147, 171
penetration tendency   172
permanent magnet   29
pH   20, 31f, 44, 49,f, 94, 127, 152, 160f, 175f
phenolic resin process   198
phosphate   27ff, 37, 39, 117, 133
phosphate addition   30
phosphate modification   27f
phosphoric acid   122
phyllosilicates   27, 63
physical solidification   23f
pinholes   199
plaster   39
plaster mould process   39
pneumatic dry reclamation   194
pneumatic reclamation   166f, 169, 171, 179
polyaddition   69
polycondensation   20, 23, 69f
polymerization   20, 44, 48, 53, 69f
polyurethane cold box process   54
pore diameters   84
Portland cement   9, 33f, 35, 73, 78ff, 85, 87
Portland slag cement   43
potash   15
potassium   14ff, 17, 20, 26, 126ff
potassium liquid glass solution   14
potassium silicate   14ff, 20
potassium silicate solution   14, 17

potassium water glass 14, 16, 126ff
potential energy 29
pouring temperature 70, 142, 151, 182, 185
powder hardener 9
Pozzolanic cement 33
precision casting procedures 9
primary strengths 27, 149, 157
process reliability 5, 9, 11
processability 56, 109
productivity 9, 11, 43, 72, 79, 81, 116, 122, 202
propylene carbonate 10, 102f, 109
PUR cold box process 38, 130, 135
pyrolysis products 119, 202, 204, 206

Q
quartz glass 132

R
radiographic examinations 63
rat tails 146
Re-use degree 164
reclaimed sand 133, 163f, 162ff, 177ff, 189ff
reclamation 8, 25, 46, 79, 107f, 110, 117, 212, 125, 133, 137, 139, 144, 163ff
reclamation degree 163f, 168, 183, 199
reclamation output A 164
recycling 12, 37, 77, 99, 107, 110, 111, 121, 124, 133, 166
refractory material 138
regeneration 38
residual strength 26ff, 30, 44f, 50, 52, 59, 66, 70f, 81, 97, 99, 102, 105, 129, 149ff, 155ff, 174, 178
resol CO_2 core sand 206
resol ester 206
road building 77
room temperature 141, 53, 59, 69, 98, 114, 118, 150, 161, 187
rotary kilns 168

S

Saab-Scania process 114
salinisation 202
salt 5, 7f, 14, 20, 37ff, 41, 49, 94, 100, 115, 117, 132f, 135ff
salt binder system 5, 7f, 37, 39, 41, 115, 132, 136
salt binding systems 38
sand temperature 63
sand waste 37f, 46
sanding 138ff
saponification 100
scanning electron microscopic images 74
scanning electron microscopy 98, 125
screen filter sets 166
segregation 49
separating systems 166
setting time 79, 84
shell mould 13, 134, 137ff, 157, 159
ship propellers 12, 88
shipbuilding 76
shootability 131
shrinkage 74
silanes 102
silanol group 20, 62
silica gel 29, 31, 48, 50, 55
silica gel drying 50
silica micelle 20f
silica sand 15f, 39, 45, 47, 50, 54f, 72, 78f, 85, 87, 90, 105, 132, 138f, 146, 148ff, 153, 157, 159, 161ff, 166ff, 182, 185, 192, 196, 204
silica sol 7ff, 12, 31, 41, 86, 136ff
silica sol binder 8, 31, 41, 137f
silica sol solutions 31
silicate anions 20, 27
silicate binder solutions 9, 17
silicate solution 9, 14ff, 23, 25f, 28ff, 36, 48, 54ff
silicic acid 9, 14, 18, 48f, 54
silicic acid micelle 18
silicon dioxide 43f
silicon-oxo-aluminate 36, 108

siloxane groups *20*
sintering behaviour *31, 147, 149, 172*
sintering phenomena *45*
sintering process *15*
sintering reaction *79*
sintering temperature *124, 146, 172, 174, 177, 189*
sintering tendency *25*
SiO_2 *14, 17f, 20f, 25, 27f, 31f, 34*
SiO_2 hydrogel *44*
SiO_2/Na_2O modulus *18, 27, 50*
SiO_4 *36*
slag cement *33, 34*
soda *15, 48f, 171*
soda water glass *31*
sodium acetate (CH_3COONa) *94ff, 101, 189, 191*
sodium bicarbonate *44, 54*
sodium carbonate *44f, 53ff, 57, 114, 157*
sodium carbonate needles *54, 97*
sodium hydrogen carbonate *53*
sodium hydroxide *16, 95, 100f, 131*
sodium ions *26, 49, 123*
sodium (Na) *14ff, 20, 26f, 54, 63, 66, 83, 102, 124, 143, 171*
sodium phosphate *27*
sodium polyphosphate *37, 132*
sodium silicate solution *18ff, 26, 28f, 55*
sodium silicates *22, 101, 112*
sodium solution *17*
sodium tripolyphosphate 6-hydrate *28*
sodium water glass *17, 30, 114, 128, 207*
sol particl *20, 101*
sol-gel conversion *21*
sol-gel process *31*
solidification *10, 20, 23f, 27f, 30, 32, 35, 37f, 41, 43, 48ff, 52ff, 62, 65, 77, 91, 94, 96f, 101f, 109, 114, 116, 126, 129, 131f, 135, 144*
solidification by drying *20f, 41*
solidification retarders *83*
solubility *38f, 50, 63, 124, 171f*
solvation *29*

solvation shell 21
Special common cement 32
specific surface area 50, 147
spiral grain crushers 166
spray-dried water glass 65
standard cement 32
starch 79, 122
steel foundries 11, 45
storage 45f, 52f, 55f, 58f, 61, 116, 119, 122f, 126ff, 133, 149, 157, 178
storage capacity 53, 65, 92, 114, 129
strength 17, 21, 23f, 32ff, 36f, 44f, 47ff, 69ff, 73f, 77ff, 94ff, 101ff, 105f, 114ff, 118ff, 140, 146, 149ff, 155ff, 161f, 169, 171f, 174, 176, 178ff, 183ff, 191ff, 196ff, 202, 204ff, 208, 210
strength values 30, 81, 118, 125, 136
stripping properties 73
sugar modification 26
sulfates 38, 136
surface quality 62, 120, 122
surface roughnes 138
surface tension 29, 138
swelling capacity 202
synthetic resin 122
synthetic sand 146

T

teflon 132
test pieces 59
thermal curing 69, 131, 157
thermal expansion 85
thermal reclamation 168, 171
thermal solidification 27
thermal stability 76
thermal stress 27, 99, 120, 147
thermo-curing inorganic binder system 56
thermoplastic behaviour 53, 102, 108, 159
three dimensional silicate network 22
titration 171, 173
titration principle 100

triacetin *90, 95, 97, 102f, 171, 192*
tricalcium aluminate *34*
tricalcium silicate *9*
tripole phosphororic acid *115*
trisodium (Na_3PO_4) *28*

U
used cement sand *193, 197f*
used sand *9, 46f, 53, 77, 89, 99, 102, 107, 110, 121, 124, 133, 163ff, 208*

V
veins *117, 146*
viscosity *14, 17ff, 28f, 31, 49, 56, 60, 62, 91, 102, 127, 131*

W
warm air drying *65*
warm air gassing *62f, 114*
warm box process *42*
warm/hot-setting moulding process *42*
waste material *11*
wastewater *169*
water glass solution *9, 14, 16, 19ff, 23f, 27ff, 36, 38, 43f, 49, 54, 56, 63, 212, 125, 136*
water extraction *61, 80*
water glass bentonite process *174*
water glass binder *5, 9f, 13f, 18, 21, 26, 31, 35, 44, 47, 49, 53, 56f, 59ff, 63, 66, 89, 91, 93, 98, 101, 107f, 114,121ff, 129, 137, 148ff, 171f, 206, 208*
water glass binder solution *26, 28, 30f, 103, 115*
water glass bonded cores *11, 24, 53*
water glass bonded moulding material *23f, 27, 44, 50, 52, 98, 114f, 131, 204*
water glass cement method *24, 79*
water glass cement process *35, 72, 89*
water glass cement processing *28*
water glass clay process *89*
water glass CO_2 *30, 174, 180, 182f, 204, 206ff*
water glass CO_2 process *5, 9, 11, 41, 168, 177, 179f, 183, 208*
water glass ester *12, 30, 90ff, 94, 97f, 102f, 105, 107ff, 130, 148f, 171ff, 177ff, 182, 187, 196*

water glass ester process   *10f, 71, 79, 88f, 94ff, 99ff, 103, 105, 107ff, 149ff, 153, 156, 160, 174, 177, 188*
water glass ferrochromium slag process   *89*
water glass ferrosilicon process   *89*
water glass film   *17, 50*
water glass hot air method   *30*
water glass hot box process   *114, 149, 151, 157f*
water glass powder curing process   *9*
water glass powder systems   *65*
water glass process   *9, 52, 58, 96f, 110, 169, 172, 178, 205*
water glass solution   *9, 14, 16, 19ff, 27f, 36, 38, 43f, 49, 54, 56, 59, 63, 121, 124, 136*
water glass warm-air process   *61, 63*
water glass CO_2 cores   *11*
water glass CO_2 process   *5, 9, 11, 41, 168, 177, 183, 208*
water glass-silicide process (Nishiyama process)   *9*
water molecules   *53, 62, 74*
water coating   *58, 119*
water glass clay process   *89*
water-soluble salt compound   *37*
water to cement ratio   *74ff, 80f*
wax pattern   *134, 137, 140*
weight losses   *63*
wet reclamation   *133, 165, 168f, 171, 183, 195ff*
wet reclamation system   *197*
wetability   *131f*
workability   *31, 45, 58, 69*

Z

Zeta-potential   *21*
zinc carbonate   *123*
zinc oxide   *122*
zircon   *138, 147, 149ff, 157*

^{29}Si NMR spectroscopy   *22, 102ff, 127*
^{31}P-NMR spectroscopy   *28*